Dynamics and Control of Robotic Manipulators with Contact and Friction

Dynamics and Control of Robotic Manipulators with Contact and Friction

Shiping Liu
Huazhong University of Science and Technology
Wuhan, Hubei, China

Gang (Sheng) Chen
Marshall University, Huntington
WV, USA

Registered Offices
John Wiley & Sons, Inc., 111 River Street, Hoboken, NJ 07030, USA
John Wiley & Sons Ltd, The Atrium, Southern Gate, Chichester, West Sussex, PO19 8SQ, UK

Editorial Office
The Atrium, Southern Gate, Chichester, West Sussex, PO19 8SQ, UK

For details of our global editorial offices, customer services, and more information about Wiley products visit us at www.wiley.com.

Wiley also publishes its books in a variety of electronic formats and by print-on-demand. Some content that appears in standard print versions of this book may not be available in other formats.

Library of Congress Cataloging-in-Publication Data

Names: Liu, Shiping, 1971- author. | Chen, Gang (Sheng), author.
Title: Dynamics and control of robotic manipulators with contact and friction
    / Dr. Shiping Liu, Huazhong University of Science and Technology,
    Wuhan, Hubei, China, Dr. Gang (Sheng) Chen, Marshall University, Huntington, WV,
    USA.
Description: First edition. | Hoboken, NJ : John Wiley & Sons, Inc., 2019. |
    Includes bibliographical references and index. |
Identifiers: LCCN 2018038795 (print) | LCCN 2018039353 (ebook) | ISBN
    9781119422495 (Adobe PDF) | ISBN 9781119422501 (ePub) | ISBN 9781119422488
    (hardcover)
Subjects: LCSH: Robots–Dynamics. | Manipulators (Mechanism)–Automatic
    control. | Tribology.
Classification: LCC TJ211.4 (ebook) | LCC TJ211.4 .L58 2018 (print) | DDC
    629.8/933–dc23
LC record available at https://lccn.loc.gov/2018038795

Cover Design: Wiley
Cover Image: © xieyuliang/Shutterstock

Set in 10/12pt WarnockPro by SPi Global, Chennai, India

Printed and bound by CPI Group (UK) Ltd, Croydon, CR0 4YY

10 9 8 7 6 5 4 3 2 1

Contents

Preface

Robotic manipulators have been widely used in the manufacturing industry. The analysis and development of robotic manipulators involves dealing with the actuation and the joint friction as well as the motions of the links in the manipulators.

To characterize friction processes and then to model and simulate them has been a challenging problem in science and engineering due to the complexity of the interface phenomenon, which involves multi-scale physics, system dynamics, operating, and environmental conditions. The dynamical systems of robotic manipulators with friction often give rise to diverse forms of complex motions, even if specified control is applied. The kinematic and dynamic relationships between the joint actuators' motion and torques, and the desired motion and force for a task, can be very complex due to existence of uncertain frictions. The design of the link and joint structures, as well as the actuation, to achieve the desired performance is accordingly challenging. The robotic manipulator is a nonlinear stochastic, coupled system that is difficult to control because of its complex dynamics.

There is no universally accepted friction model or theory to cover general friction phenomena due to its multi-physics nature. A different model has been developed for individual conditions. Part of the reason is due to the fact that friction is a complex process in which forces are transmitted, mechanical energy is converted, surface topography is altered, interface material could be removed or formed, and physical and even chemical change could occur. Actually, friction could be assumed as a variable in a dynamical system having sliding interface, which is complicated by the fact that this dynamical system's boundary condition is not stationary and deterministic due to interfacial changes in geometrical, mechanical, material, physical, and chemical aspects. As such, the coefficients of friction are not intrinsic properties of materials. They depend on the properties of the contact surfaces, their operational conditions, their time history, environmental conditions, and even their interactions.

This book presents the fundamental principles and methods that are used to develop a robotic manipulator system. To give some examples of the problems treated in the book, let's consider the immense efforts that are being put into dealing with friction dynamics of robotic manipulators. In the modeling of robotic manipulator to realize the precise motion control, friction compensation is a crucial step.

Friction effects quantification are particularly critical for industrial robotic manipulators analysis and control. It has been observed that friction can cause 50% error in some heavy industrial manipulators. A poor friction compensation action in the control scheme may lead to significant tracking errors, stick-slip motions, hunting in the stopping phase of the robot movement, and limit cycles when velocity reversals occur in the assigned trajectory.

In the past decades, the applications of robot manipulators have also widely extended to the area of healthcare in hospitals/patient-care/ surgery. The medical robot is a new trend in medicine, which aims to help the surgeon by taking advantage of high accuracy and accessibility of robots. This offers advantages including off-loading of routine tasks and reduction of the number of human assistants in the hospital operating room. The surgeons can complement their own skills with the accuracy, motion steadiness, and repeatability of the robot. In most of these surgical operations, robots serve as an aid or as an extension of the doctor's capabilities.

For example, a famous surgical system is a successfully commercialized robotic surgical system used in hospitals worldwide, which was designed to facilitate complex surgery using a minimally invasive approach, and is controlled by a surgeon from a console. The system is commonly used for prostatectomies, and increasingly for cardiac valve repair and gynecologic surgical procedures. The system is a remote-control robot featuring four manipulators that are controlled by a surgeon through the use of hand and foot controls while sitting at a console that provides a virtual-reality representation of the patient's internal organs. The surgical robot has been increasingly used at hospitals for a number of different urologic, laparoscopic, gynecologic, and noncardiovascular thoracoscopic surgical procedures and thoracoscopically assisted cardiotomy procedures. However, it was once reported that the supplier company issued a recall affecting more than 1000 robot arms around the world because they might be producing too much friction in some of the surgical systems. When that happens, the surgical system can choke during surgery and briefly stop working before it catches up. Considering that the system costs more than $1.5 million per unit, and that the recall affected more than a thousand systems globally, we can get a sense of how serious the problem is. According to the reports, friction within the instrument arms of the surgical system could interrupt the system arm movement, resulting in them stalling or getting stuck. The problem resulted in an imprecise cut during robotic surgery. FDA recorded the manufacturer's reason for recall of the patient side manipulator on the surgical systems: friction within certain instrument arms can interrupt smooth instrument motion. This can be felt by the surgeon as resistance in the movement of the master. In this situation, the instrument can stall momentarily and then suddenly catch up to the master position if the surgeon pushes through the resistance.

Understanding the nature of friction dynamics and solving the technological problems associated with the friction dynamics of manipulators are the essence of these fields. Modeling of friction dynamics in manipulator systems requires an accurate description of friction. Unfortunately, there is no universally accepted friction model or theory to cover general friction phenomena. The resultant dynamics of manipulators often exhibit various nonlinear, transient/nonstationary, and uncertain properties. Moreover, small changes in interfacial parameters could have a significant effect on the resultant dynamics, and thus the scales of influencing factors span from macro-, micro-, to nano-meter level. The boundary condition of the problems is not fixed or given beforehand; actually, it depends on environmental conditions, operation conditions, system interactions, and time.

Because of the complexity of the friction dynamics of manipulators, it has been considered to be an unsolved problem in many robotic engineering applications. Since the modeling and predictions are not very reliable, the "trial and error" approach has been extensively used. The recent extensive efforts on modeling, analytical, and experimental investigation have made substantial progress in many practical applications. Many techniques, such as advanced testing techniques, advanced signal processing techniques, and spectrum analysis and contemporary nonlinear dynamics techniques, as well as advanced control technologies, have been used as efficient means to address the nonlinear, nonstationary, and uncertain dynamics. These enable researchers to efficiently quantify the friction dynamics of manipulators. The emergence and applications of IT-based approaches have allowed systematic solutions to be possible.

Research in friction dynamics of manipulators has several purposes. To name a few: to develop fundamental understanding of friction dynamics of manipulators, to control the motion of dynamical system of robotic manipulators with friction; to realize some physical processes for products of robots; and, understandably, to reduce and eliminate the instability in robotic manipulator systems caused by friction.

This book introduces the basic concepts and principles of friction dynamics of manipulators and controls. This book offers a combined treatment of the modeling, analysis, and control of many manipulator problems. After delineating these mathematical characterizations, it presents many applications in use today for analyzing friction dynamics of manipulators. Emphasis is on the fundamental aspects and the contemporary knowledge of the dynamics and control of robotic manipulator with contact and friction.

1

Introduction

1.1 Robot Joint Friction Modeling and Parameter Identification

Robot design generally considers only the ideal circumstances. The actual robot experiences great differences from the model built and simulated with MATLAB or other graphical tools, with manufacturing errors, friction, and gravity contributing the most to these differences [1]. There are also complex nonlinear problems [2]. With the wide application of robots in different industries involving humans, the demands of safety, accuracy, and reliability on robots is continually increasing. Contact and friction always influence the accuracy and reliability of the robot. For a more reliable and accurate solid robot, the design – especially control design – must account for friction, contact, and impact, as well as friction in the force feedback mechanism.

In recent decades, industries and academia have conducted research and developed related to robotic dynamics and control. In recent years, some have focused on the contact and friction generated in robot joints during movement [3–8]. There have been different methods presented to build an accurate model of joint friction based on experimentation and analysis of the corresponding results. Nonlinear characteristics of friction [9–15] are obvious, especially in the low-speed motion of robot, so nonlinear dynamics of robot must be considered as well.

Friction is a tangential resistant force to sliding in a dynamic system. To investigate friction, it is necessary to characterize the surface roughness in terms of its statistical properties. Friction has been demonstrated to be related to surface topography, and friction

Dynamics and Control of Robotic Manipulators with Contact and Friction, First Edition.
Shiping Liu and Gang (Sheng) Chen.
© 2019 John Wiley & Sons Ltd. Published 2019 by John Wiley & Sons Ltd.

investigation has been conventionally attributed to the determination of the actual area of contact and understanding the contact mechanism. For instance, surface physics explains friction as the formation of adhesion between interacting asperities and their breakaway by shearing, whereas continuum mechanics interprets friction by interlocking and subsequent fracture of asperities.

To address the problem of actual sliding asperity contact is quite difficult, which involves complex dynamics. The feasible approach is to assume the contact to be quasi-static in nature, to certain extent. In many applications with relatively smooth surfaces, the deformation of contacting asperities can be assumed to be linear and elastic. For many problems the contact has to be extended to non-elastic and nonlinear conditions and involve dynamics.

1.2   Contact Perception in Virtual Environment

Research on robot dynamics and control with consideration of contact and friction is a major topic of this book. Contact perception, which is part of the interaction between a human and a real or virtual robot, is also discussed. In a virtual environment, force feedback or surface texture generally needs to be calculated according to dynamics with friction and contact.

Contact perception is generally achieved by different types of haptic devices [16–22]. The force signals or the surface texture signals, coming from the real measurement or calculation, is input into the haptic device so the operator gets the expected contact feeling. The real measurement is carried out when the real robot works in the real environment. In virtual simulation, the virtual interaction force between the virtual robot and the virtual environment needs to be calculated in real time according to relevant contact or friction models. The measured or calculated force signals are filtered and sometimes transformed according to a certain scale.

Virtual reality and mixed reality are widely used in robot control, especially for control simulation [23–31]. The purpose of the virtual simulation in some instances is to verify the underlying robot control algorithm, while others are to provide graphical modes to assist the operator with real-time robot control, reducing manual error.

This book mainly discusses the related research of virtual reality or augmented reality teleoperation [30–37]. Interactive task planning with graphical assistance is also explored.

1.3 Organization of This Book

Chapter 1 introduces the book and major topics are stated.

Chapter 2 introduces the fundamentals of robot dynamics and control. Besides the typical kinematics and dynamics of a robot with six degrees of freedom (6 DoF) or less, the kinematic reverse solution of a 7 DoF robot is also introduced. In this chapter, different robot control modes are classified as either trajectory control or interaction control, and reviewed.

In Chapter 3, theories and methods in contact and friction are reviewed according to the classification of wet friction and dry friction. In this chapter, we present the fundamentals of contact and friction between two contact surfaces in the context of quasi-static state by assuming that the normal motion is ignored. We focus on the mechanics of contact and friction by outlining the mechanical attributes of various friction processes in the context of the problems of the friction-vibration interactions.

Chapter 4 introduces friction modeling and parameter identification of robot joints. The dynamic parameter identification methods of multiple-joint robot systems are also introduced. This chapter uses the two-link planar robotic arm as the experimental object to verify theories and methods discussed in previous chapters. Nonlinear dynamics and chaos are also discussed in this chapter.

The operator of a remote or virtual robot can feel the contact between the robot and its real or virtual working environment via a haptic device. In Chapter 5, principles of several common haptic devices with force feedback are analyzed. The calculation of virtual force caused by contact in a virtual environment is discussed. Haptic display based on point haptic devices is also reviewed.

Chapter 6 introduces virtual simulation of robot control and 3D graphic environment, virtual reality–based robot control and augmented reality–based teleoperation are reviewed. Task planning based on graphical mode is also discussed.

References

1 Japan Robot Association (2008). *The New Handbook of Robot Technology* Zong G.H., Cheng J.S., trans., 2e, 244–245. Beijing: Science Press.

2 Verduzco, F. and Alvarez, J. (2000). Homoclinic chaos in 2-DOF robot manipulators driven by PD controllers. *Nonlinear Dynam.* 21 (2): 157–171.

3 Parra-Vega, V. and Arimoto, S. (1996). A passivity based adaptive sliding mode position-force control for robot manipulators. *Int. J. Adapt. Control Signal Process.* 10 (4–5): 365–377.

4 de Wit, C.C., Olsson, H., Astron, K.J. et al. (1994). A new model for control of systems with friction. *IEEE Trans. Autom. Control* 40: 419–425.

5 Popovic, M.R., Gorinevsky, D.M., and Goldenberg, A.A. (2000). High-precision positioning of a mechanism with nonlinear friction using a fuzzy logic pulse controller. *IEEE Trans. Control Syst. Technol.* 8 (1): 151–158.

6 Llama, M.A., Kelly, R., and Santibáñez, V. (2000). Stable computedtorque control of robot manipulators via fuzzy self-tuning. *IEEE Trans. Syst., Man, Cybern. – Part B* 30 (1): 143–150.

7 Craig, J.J. (1988). *Adaptive Control of Mechanical Manipulators.* New York: Addison-Wesley Inc.

8 Wu, J., Wang, J., and You, Z. (2010). An overview of dynamic parameter identification of robots. *Robot. Comput. Integr. Manuf.* 26 (5): 414–419.

9 Wang, S.-G., Lin, S.B., Shieh, L.S. et al. (1998). Observer-based controller for robust pole clustering in a vertical strip and disturbance rejection in structured uncertain systems. *Int. J. Robust Nonlinear Control* 8 (3): 1073–1084.

10 Lin, S. and Wang, S.-G. (2000). Robust control with pole clustering for uncertain robotic systems. *Int. J. Control Intell. Syst.* 28 (2): 72–79.

11 Ryu, J.-H., Song, J., and Kwon, D.-S. (2001). A nonlinear friction compensation method using adaptive control and its practical application to an in-parallel actuated 6-DOF manipulator. *Control. Eng. Pract.* 9: 159–167.

12 Jin, M., Kang, S.H., and Chang, P.H. (2008). Robust compliant motion control of robot with nonlinear friction using time-delay estimation. *IEEE Trans. Ind. Electron.* 55 (1): 258–269.

13 Chang, P.H., Park, K., Kang, S.H. et al. (2013). Stochastic estimation of human arm impedance using robots with nonlinear frictions: an experimental validation. *IEEE/ASME Trans. Mechatron.* 18 (2): 775–786.

14 Sneider, H. and Frank, P.M. (1996). Observer-based supervision and fault detection in robots using nonlinear and fuzzy logic residual evaluation. *IEEE Trans. Control Syst. Technol.* 4 (3): 274–282.

15 Do, T.N., Tjahjowidodo, T., Lau, M.W.S. et al. (2015). Nonlinear friction modelling and compensation control of hysteresis phenomena for a pair of tendon-sheath actuated surgical robots. *Mech. Syst. Signal Process.* 60–61: 770–784.

16 Salisbury, K., Conti, F., and Barbagli, F. (2004). Haptic rendering: introductory concepts. *IEEE Comput. Graphics Appl.* 24 (2): 24–32.

17 Yi, L., Zhang, Y., Ye, X. et al. (2016). Haptic rendering method based on generalized penetration depth computation. *Signal Process.* 120: 714–720.

18 Miguel Angel Otaduy Tristan (2004). *6-DOF Haptic Rendering Using Contact Levels of Detail and Haptic Textures*. Chapel Hill: Dissertation of the University of North Carolina.

19 Okamoto, S., Konyo, M., Saga, S. et al. (2009). Detectability and perceptual consequences of delayed feedback in a vibrotactile texture display. *IEEE Transactions on Haptics* 2 (2): 73–84.

20 CB Zilles and J.K. Salisbury, 1995 A Constraint-Based God-Object Method for Haptic Display. Proc. IEE/RSJ Int'l Conf. Intelligent Robots and Systems, Human Robot Interaction, and Cooperative Robots, vol. 3, IEEE CS Press, pp. 146–151.

21 Minsky, M. (1995). Computational Haptics: The Sandpaper System for Synthesizing Texture for a Force Feedback Display, doctoral dissertation, Mass. Inst. of Technology.

22 Juan, W., Ju, Y., Li-yuan, L. et al. (2013). Design and implementation of measurement-based texture force rendering. *J. Syst Simul.* 25 (11): 2630–2636.

23 Sud, A., Andersen, E., Curtis, S. et al. (2008). Real-time path planning in dynamic virtual environments using multiagent navigation graphs. *IEEE Trans.Visual. Comput. Graphics* 14 (3): 526–538.

24 Erez, T. Tassa, Y. and Todorov, E. (2015). Simulation Tools for Model-Based Robotics: Comparison of Bullet, Havok, MuJoCo, ODE and PhysX. IEEE International Conference on Robotics and Automation (ICRA).

25 Brooks, F.P. (1999). What's real about virtual reality? *IEEE Comput. Graph. Appl.* 19 (6): 16.

26 Earnshaw, R.A. (ed.) (2014). *Virtual Reality Systems*. Cambridge, MA: Academic Press.

27 NASA. (n.d.). Station Spacewalk Game. https://www.nasa.gov/multi media/3d resources/station spacewalk game.html

28 Qin Ping, Z. (2009). A survey on virtual reality. *Sci. China Ser. F: Inf. Sci.* 52 (3): 348–400.

29 Bowyer, S.A., Davies, B.L., and y Baena, F.R. (2014). Active constraints/virtual fixtures: a survey. *IEEE Trans. Robot.* 30 (1): 138–157.

30 Chintamani, K., Cao, A., Darin Ellis, R. et al. (2010). Improved telemanipulator navigation during display-control misalignments using augmented reality cues. *IEEE Trans. Syst. Man Cybern. Syst. Hum.* 40 (1): 29–39.

31 Yamamoto, T., Abolhassani, N., Jung, S. et al. (2012). Augmented reality and haptic interfaces for robot-assisted surgery. *Int. J. Med. Rob. Comput. Assisted Surg.* 8: 45–56.

32 Fong, T. and Thorpe, C. Vehicle teleoperation interfaces. *Auton. Robot.* 11 (1): 9–18.

33 Passenberg, C., Peer, A., and Buss, M. (2010). A survey of environment-, operator-, and task-adapted controllers for teleoperation systems. *Mechatronics* 20 (7): 787–801.

34 Zainan, J., Hong, L., Jie, W. et al. (2009). Virtual reality-based teleoperation with robustness against modeling errors. *Chin. J. Aeronaut.* 22: 325–333.

35 Fusiello, A. and Murino, V. (2004). Augmented scene modeling and visualization by optical and acoustic sensor integration. *IEEE Trans. Vis. Comput. Graph.* 10 (6): 625–636.

36 Portilla, H. and Basañez, L. (2007). Augmented reality tools for enhanced robotics teleoperation systems. 3DTV Conference, Kos Island, pp. 1–4.

37 Leutert, F. and Schilling K. (2012). Support of Power Plant Tele-maintenance with Robots by Augmented Reality Methods. 2nd International Conference on Applied Robotics for the Power Industry (CARPI) ETH Zurich, Switzerland. 11–13: 45–49.

2

Fundamentals of Robot Dynamics and Control

2.1 Robot Kinematics

Robot kinematics is classified into two categories: forward kinematics and inverse kinematics. Forward kinematics is used to determine the posture of the robot's hand based on the of displacement or velocity inputs from each joint. Inverse kinematics is used to calculate the value of each joint variable (its angular displacement, linear displacement, or velocity) given the specific point and gesture of the robot end. A matrix is used to establish the representation method of object position, object gesture and object motion, and then the forward and inverse kinematics of different robot configurations, such as Cartesian coordinates, cylindrical coordinates, and spherical coordinates are studied. The Denavit-Hartenberg (D-H) method [1] is used to derive the forward and inverse kinematic equations for all possible robot configurations.

2.1.1 Matrix Description of Robot Kinematics

Matrices can be used to represent points, vectors, coordinate systems, translations, rotations, and transformations, and can represent objects and other moving elements in the coordinate system.

The spatial point P (shown in Figure 2.1) can be represented by its three coordinates relative to the reference coordinate system:

$$P = a_x i + b_y j + c_z k \tag{2.1}$$

where (a_x, b_y, c_z) is the coordinate of that point in the reference coordinate system.

Dynamics and Control of Robotic Manipulators with Contact and Friction, First Edition.
Shiping Liu and Gang (Sheng) Chen.
© 2019 John Wiley & Sons Ltd. Published 2019 by John Wiley & Sons Ltd.

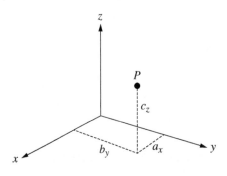

Figure 2.1 The represent of spatial point P.

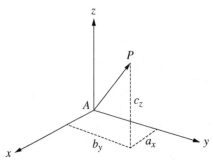

Figure 2.2 The representation of spatial vector P.

A vector begins from one spatial point and ends at another spatial point. If the beginning point is A and the ending point is B, it can be represented as:

$$\overline{P}_{AB} = (B_x - A_x)i + (B_y - A_y)j + (B_z - A_z)k \tag{2.2}$$

If A is the origin, as shown in Figure 2.2, then

$$\overline{P} = a_x i + b_y j + c_z k \tag{2.3}$$

The three components of the vector can be written in matrix form:

$$\overline{P} = \begin{bmatrix} a_x \\ b_y \\ c_z \end{bmatrix} \tag{2.4}$$

For convenience of further matrix calculations, this matrix can be expanded as:

$$\overline{P} = \begin{bmatrix} a_x \\ b_y \\ c_z \\ 1 \end{bmatrix} \tag{2.5}$$

Figure 2.3 Representation of coordinate frame in reference coordinate frame.

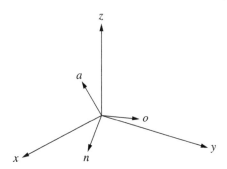

A coordinate frame in which its origin is on the origin of the reference coordinate frame is generally represented by three unit vectors, $\bar{n}, \bar{o}, \bar{a}$, which are perpendicular to each other and represent normal, orientation, and approach vectors, as shown in Figure 2.3. Each unit vector is represented by three components in the reference frame. So the coordinate frame can be represented as follows:

$$F = \begin{bmatrix} n_x & o_x & a_x \\ n_y & o_y & a_y \\ n_z & o_z & a_z \end{bmatrix} \tag{2.6}$$

This matrix generally is expanded as:

$$F = \begin{bmatrix} n_x & o_x & a_x & 0 \\ n_y & o_y & a_y & 0 \\ n_z & o_z & a_z & 0 \\ 0 & 0 & 0 & 1 \end{bmatrix} \tag{2.7}$$

The representation of a rigid object in space can be achieved by affixing a coordinate frame onto it, as shown in Figure 2.4. The position

Figure 2.4 The representation of a spatial rigid object.

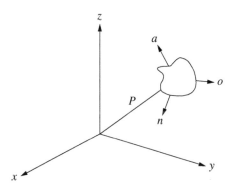

of the object relative to the coordinate frame is known. The spatial coordinate frame can be represented by a matrix, as just described. The vector from the origin of the reference matrix to the attached coordinate frame is described as:

$$\overline{P} = \begin{bmatrix} p_x \\ p_y \\ p_z \\ 1 \end{bmatrix} \tag{2.8}$$

So, the position and gesture of a rigid object can be described as:

$$F_{object} = \begin{bmatrix} n_x & o_x & a_x & p_x \\ n_y & o_y & a_y & p_y \\ n_z & o_z & a_z & p_z \\ 0 & 0 & 0 & 1 \end{bmatrix} \tag{2.9}$$

In this matrix there are constraints, such as: $\overline{n} \cdot \overline{o} = 0$, $\overline{n} \cdot \overline{a} = 0$, $\overline{a} \cdot \overline{o} = 0$, $|n| = 1$, $|o| = 1$, and $|a| = 1$.

2.1.2 Homogeneous Transformation Matrices

Robots have either translational or rotational joints. So, a unified mathematical description of translational and rotational displacements is needed [2].

The translational displacement $\overline{d}$ in Figure 2.5, given by the vector

$$\mathbf{d} = d_x \mathbf{i} + d_y \mathbf{j} + d_z \mathbf{k},$$

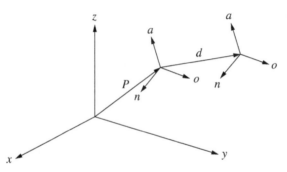

Figure 2.5 The representation of translational transform.

can also be described by the following homogeneous transformation matrix **T**:

$$
\mathbf{T} = \begin{bmatrix} 1 & 0 & 0 & d_x \\ 0 & 1 & 0 & d_y \\ 0 & 0 & 1 & d_z \\ 0 & 0 & 0 & 1 \end{bmatrix}
\tag{2.10}
$$

where (d_x, d_y, d_z) are the three components of the transitional vector $\overline{d}$ along three axes x, y, z in the reference frame. So, the new position of the coordinate frame is described as:

$$
F_{new} = \begin{bmatrix} 1 & 0 & 0 & d_x \\ 0 & 1 & 0 & d_y \\ 0 & 0 & 1 & d_z \\ 0 & 0 & 0 & 1 \end{bmatrix} \times \begin{bmatrix} n_x & o_x & a_x & p_x \\ n_y & o_y & a_y & p_y \\ n_z & o_z & a_z & p_z \\ 0 & 0 & 0 & 1 \end{bmatrix} = \begin{bmatrix} n_x & o_x & a_x & p_x + d_x \\ n_y & o_y & a_y & p_y + d_y \\ n_z & o_z & a_z & p_z + d_z \\ 0 & 0 & 0 & 1 \end{bmatrix}
\tag{2.11}
$$

It can also be in the form:

$$
F_{new} = Trans(d_x, d_y, d_z) \times F_{old}
\tag{2.12}
$$

In rotational transformation, the right-hand rule is used, in which the thumb is directed along the axis toward its positive end, while the fingers show the positive direction of the rotational displacement. Generally, we assume that the origin of the coordinate frame to be rotated is on the origin of the reference coordinate frame and all their three axes are coincident, respectively.

If we only rotate the coordinate frame around the x-axis, as shown in Figure 2.6, the rotational matrix is given as:

$$
\begin{bmatrix} P_x \\ P_y \\ P_z \\ 1 \end{bmatrix} = \begin{bmatrix} 1 & 0 & 0 & 0 \\ 0 & \cos\theta & -\sin\theta & 0 \\ 0 & \sin\theta & \cos\theta & 0 \\ 0 & 0 & 0 & 1 \end{bmatrix} \begin{bmatrix} P_n \\ P_o \\ P_a \\ 1 \end{bmatrix}
\tag{2.13}
$$

The rotational matrix can also be written in this form:

$$
P_{xyz} = Rot(x, \theta) \times P_{noa}
\tag{2.14}
$$

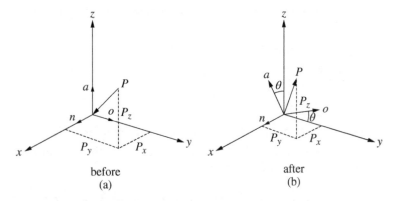

before
(a)

after
(b)

Figure 2.6 Coordinates of point before and after the rotation of frame.

For simplification, $C\theta$ and $S\theta$ are used to denote $\cos\theta$ and $\sin\theta$, respectively. So, the rotation matrix can also can be written as:

$$Rot(x, \theta) = \begin{bmatrix} 1 & 0 & 0 & 0 \\ 0 & C\theta & -S\theta & 0 \\ 0 & S\theta & C\theta & 0 \\ 0 & 0 & 0 & 1 \end{bmatrix} \tag{2.15}$$

Similarly, the rotation matrices around the y-axis and the z-axis are written as below:

$$Rot(y, \theta) = \begin{bmatrix} C\theta & 0 & S\theta & 0 \\ 0 & 1 & 0 & 0 \\ -S\theta & 0 & C\theta & 0 \\ 0 & 0 & 0 & 1 \end{bmatrix} \text{ and } Rot(z, \theta) = \begin{bmatrix} C\theta & -S\theta & 0 & 0 \\ S\theta & C\theta & 0 & 0 \\ 0 & 0 & 1 & 0 \\ 0 & 0 & 0 & 1 \end{bmatrix} \tag{2.16}$$

The rotation around an arbitrary axis passing through the origin can be considered as a series of rotations about these three coordinate axes. Here, the sequence is important; if the sequence is different, the final result will be completely different.

2.1.3 Forward Kinematics

The D-H convention has been the standard method used to model the robot and its movement. It can apply to any robot configuration no matter the structural sequence and complexity.

Think of the composition of robots as a series of links and joints. These joints can be prismatic (linear) or revolute (hinged), which can be placed in any order and in any plane. The links can be of any length (including zero), may be bent or twisted, and may be on any plane. So any set of joints and links can form a robot we want to model and represent.

First, a reference coordinate frame is specified for each joint, and then the steps to transform from one joint to the next joint (one coordinate frame to the next coordinate frame) is determined. The total transformation matrix of the robot is obtained by sequentially combining all the transformations from the base to the last joint.

In Figure 2.7, three sequential joints and two links in a robot are shown. These joints may be revolute, prismatic, or both. Although the joints of the robot are usually only one degree of freedom (DoF), the joints in Figure 2.7 may sometimes represent two degrees of freedom.

The first joint is designated as joint n, the second joint is joint $n + 1$, and the third joint is joint $n + 2$. There may be other joints before and after these joints. The link n is between the joints n and $n + 1$ and the link $n + 1$ is between the joints $n + 1$ and $n + 2$.

For each joint to establish a reference coordinate frame, a z-axis and x-axis must be specified. Usually there is no need to specify the y-axis because the y-axis is always perpendicular to the x-axis and z-axis. The following is the procedure for assigning a local reference frame to each joint:

- All joints, without exception, are the z-axis. If the joint is rotated, the z-axis is in the direction of rotation by the right-hand rule. If the joint is prismatic, the z-axis is in the direction of the straight movement. In each case, the subscript of the z-axis (and the local reference coordinate frame of the joint) at the joint n is $n - 1$. For example, the z-axis representing the number of joints $n + 1$ is n. These simple rules allow us to quickly define the z-axis of all joints. For revolute joints, the angular displacement θ around the z-axis is the joint variable. For a prismatic joint, the link length d along the z-axis is a joint variable.

- As shown in Figure 2.7, joints are generally not parallel or intersecting. Thus, adjacent z-axes are skewed, but there is always a shortest vertical line, which is orthogonal to both of these two skew lines. The x-axis of the local reference coordinate frame is usually defined in the vertical direction. So if the vertical line between z_{n-1} and z_n is

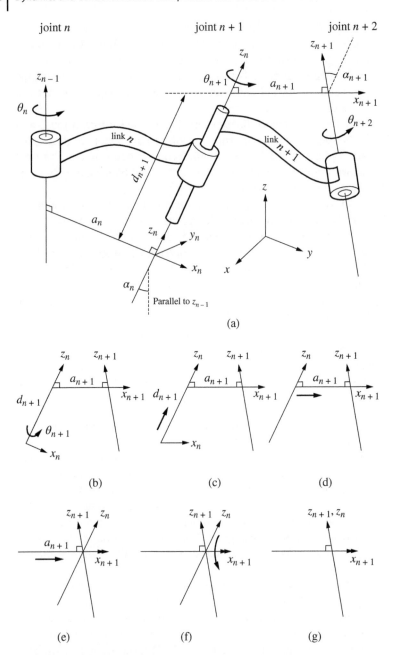

Figure 2.7 D-H representation of combination of joints and links.

represented with the line a_n, then the direction of x_n will go along a_n. Likewise, the vertical line between z_n and z_{n+1} is represented with the line a_{n+1}, the direction of x_{n+1} will go along a_{n+1}. Note that the vertical lines between adjacent joints do not necessarily intersect and are not collinear, so, the positions of the two adjacent coordinate frame origins may not be in the same position. Based on the information described above and taking into account the special circumstances of the following exceptions, the coordinate frames for all joints can be defined.

- If the z-axes of the two joints are parallel, there are numerous vertical lines between them. In this case, a vertical line can be selected that is collinear with the vertical line of the previous joint, which simplifies the model.

- If the z-axes of two adjacent joints are intersecting, there is no vertical line between them (or the vertical distance is zero). In this case, a straight line perpendicular to the plane formed by the two axes can be defined as the x-axis. In other words, its vertical line is perpendicular to the plane containing two z-axes. It is also equivalent to select the two z-axes cross product direction as the x-axis. This will also simplify the model.

In Figure 2.7a, the angle θ represents the rotation angle around the z-axis, d represents the distance between two adjacent vertical lines on the z-axis, a represents the length of each vertical line (also called the joint displacement), and the angle α represents the angle between two adjacent z-axes (also called joint twist). Usually, only θ and d are joint variables.

Next, a reference coordinate frame is transformed to the next reference coordinate frame. Assume that the current reference coordinate frame is $x_n - z_n$, then the next reference coordinate frame $x_{n+1} - z_{n+1}$ can be attained through the following four standard movements.

1) Rotate θ_{n+1} around the axis z_n, (as shown in Figures 2.7a,b), to make x_n and x_{n+1} be parallel to each other.
2) Move d_{n+1} along z_n to make x_n and x_{n+1} collinear, (as shown in Figure 2.7c).
3) Move a_{n+1} along x_n, to make origins of x_n and x_{n+1} coincide (as shown in Figure 2.7d,e).
4) Rotate z_n about α_{n+1} about x_{n+1}, to make z_n and z_{n+1} collinear (as shown in Figure 2.7f). Now the coordinate frame n and $n+1$ become the same (as shown in Figure 2.7g).

The coordinate frame transformation can be made between coordinate frame $n+1$ and $n+2$ strictly in accordance with the same four orders of motion. Repeating the above steps, transformation of a series of adjacent coordinate frames can be achieved. Starting from the reference coordinate frame, we can convert it to the base of the robot, and then to the first joint, the second joint, etc. until the end effector. The benefit is that the transformation between any two coordinate systems can be the same as before.

The transformation matrix A is obtained by the right multiplying the four matrices representing the four movements, and the matrix A represents four successive movements. Since all transformations are relative to the current coordinate frame (i.e. they are measured and executed relative to the current local coordinate frame), all matrices are right multiplied. The result is as follows:

$$^{n}T_{n+1} = A_{n+1} = Rot(z, \theta_{n+1}) \times Tran(0, ..., 0, ..., d_{n+1})$$
$$\times Tran(a_{n+1}, 0, 0) \times Rot(x, \alpha_{n+1})$$

$$= \begin{bmatrix} C\theta_{n+1} & -S\theta_{n+1} & 0 & 0 \\ S\theta_{n+1} & C\theta_{n+1} & 0 & 0 \\ 0 & 0 & 1 & 0 \\ 0 & 0 & 0 & 1 \end{bmatrix} \times \begin{bmatrix} 1 & 0 & 0 & 0 \\ 0 & 1 & 0 & 0 \\ 0 & 0 & 1 & d_{n+1} \\ 0 & 0 & 0 & 1 \end{bmatrix}$$

$$\times \begin{bmatrix} 1 & 0 & 0 & a_{n+1} \\ 0 & 1 & 0 & 0 \\ 0 & 0 & 1 & 0 \\ 0 & 0 & 0 & 1 \end{bmatrix} \times \begin{bmatrix} 1 & 0 & 0 & 0 \\ 0 & C\alpha_{n+1} & -S\alpha_{n+1} & 0 \\ 0 & S\alpha_{n+1} & C\alpha_{n+1} & 0 \\ 0 & 0 & 0 & 1 \end{bmatrix} \tag{2.17}$$

$$A_{n+1} = \begin{bmatrix} C\theta_{n+1} & -S\theta_{n+1}C\alpha_{n+1} & S\theta_{n+1}S\alpha_{n+1} & a_{n+1}C\theta_{n+1} \\ S\theta_{n+1} & C\theta_{n+1}C\alpha_{n+1} & -C\theta_{n+1}S\alpha_{n+1} & a_{n+1}S\theta_{n+1} \\ 0 & S\alpha_{n+1} & C\alpha_{n+1} & d_{n+1} \\ 0 & 0 & 0 & 1 \end{bmatrix} \tag{2.18}$$

On the base of the robot, transformation from the first joint to the second joint can be done, then to the third, etc., to the last joint of the robot. If each transformation is defined as $^{n}T_{n+1}$, then a number of transformations can be obtained. The total conversion between the base and the last robot joint is:

$$^{R}T_{H} = {}^{R}T_{1}\,{}^{1}T_{2}\,{}^{2}T_{3} \cdots {}^{n-1}T_{n} = A_{1}A_{2}A_{3} \cdots A_{n} \tag{2.19}$$

where n is total number of joints.

Table 2.1 D-H parameters.

#	a	α	d	θ
1				
2				
3				
4				
5				
6				

To simplify the calculation of the A matrix, a table of joint and link parameters can be made, where the parameter values for each link and joint can be determined from the schematic diagram of the robot and these parameters can be substituted into the A matrix. Table 2.1 can be used for this purpose.

Next we take the robot shown in Figure 2.8 as an example to analyze its forward kinematics.

When the robot does not work, its pose is shown in Figure 2.8. This pose is set as the robot zero pose. In this pose, the big arm is vertical, the small arm is horizontal, and the three axes of the wrist are on the same plane, which is parallel to the big and small arm, and through the center of the base.

According to D-H convention, coordinate frames are built as shown in Figure 2.9. In that figure, the axes that are perpendicular

Figure 2.8 A robot with 6 DOF [3].

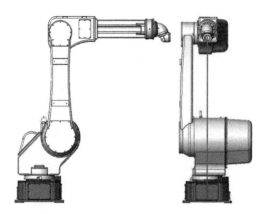

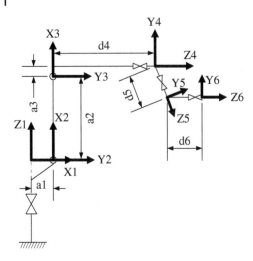

Figure 2.9 Coordinate frame for each joint.

Table 2.2 The D-H parameters of the robot.

Link i	a_{i-1}	α_{i-1}	d_i	θ_i
1	0	0	0	Variable
2	270	−90°	0	Variable
3	1300	0	0	Variable
4	42.5	−90°	1300	Variable
5	0	70°	108.9	Variable
6	0	−70°	82	Variable

to the paper are not drawn, and they are decided by right-hand rule. Table 2.2 lists the D-H parameters decided by D-H coordinate frames. The initial values for joint angles are $\theta_1 = \theta_3 = \theta_5 = \theta_6 = 0$ and $\theta_2 = \theta_4 = -90°$.

It should be noted that the base coordinate frame $x_0 - z_0$ should be fixed on the base (this is not shown in Figure 2.9). Theoretically, the base coordinate frame as the world coordinate frame can be fixed arbitrarily, but for the convenience of calculation, it is generally coincident with the coordinate system $x_1 - z_1$ when the first joint is in the initial state.

Substituting the corresponding D-H parameters into Eq. (2.18) yields the link transformation matrix as follows:

$$
{}_1^0T = \begin{bmatrix} c_1 & -s_1 & 0 & 0 \\ s_1 & c_1 & 0 & 0 \\ 0 & 0 & 1 & 0 \\ 0 & 0 & 0 & 1 \end{bmatrix}; \quad
{}_2^1T = \begin{bmatrix} c_2 & -s_2 & 0 & a_1 \\ 0 & 0 & 1 & 0 \\ -s_2 & -c_2 & 0 & 0 \\ 0 & 0 & 0 & 1 \end{bmatrix}
$$

$$
{}_3^2T = \begin{bmatrix} c_3 & -s_3 & 0 & a_2 \\ s_3 & c_3 & 0 & 0 \\ 0 & 0 & 1 & 0 \\ 0 & 0 & 0 & 1 \end{bmatrix}; \quad
{}_4^3T = \begin{bmatrix} c_4 & -s_4 & 0 & a_3 \\ 0 & 0 & 1 & d_4 \\ -s_4 & -c_4 & 0 & 0 \\ 0 & 0 & 0 & 1 \end{bmatrix}
$$

$$
{}_5^4T = \begin{bmatrix} c_5 & -s_5 & 0 & 0 \\ s_5 c\alpha_4 & c_5 c\alpha_4 & -s\alpha_4 & -d_5 s\alpha_4 \\ s_5 s\alpha_4 & c_5 s\alpha_4 & c\alpha_4 & d_5 c\alpha_4 \\ 0 & 0 & 0 & 1 \end{bmatrix};
$$

$$
{}_6^5T = \begin{bmatrix} c_6 & -s_6 & 0 & 0 \\ s_6 c\alpha_5 & c_6 c\alpha_5 & -s\alpha_5 & -d_6 s\alpha_5 \\ s_6 s\alpha_5 & c_6 s\alpha_5 & c\alpha_5 & d_6 c\alpha_5 \\ 0 & 0 & 0 & 1 \end{bmatrix} \tag{2.20}
$$

By multiplying these link transformation matrices, we can get the transformation matrix of the robot end coordinate frame to the base frame ${}_6^0T$:

$$
{}_6^0T = {}_1^0T\,{}_2^1T\,{}_3^2T\,{}_4^3T\,{}_5^4T\,{}_6^5T = \begin{bmatrix} n_x & o_x & a_x & p_x \\ n_y & o_y & a_y & p_y \\ n_z & o_z & a_z & p_z \\ 0 & 0 & 0 & 1 \end{bmatrix} \tag{2.21}
$$

In the following expressions, s_{23} denotes $\sin(\theta_2 + \theta_3)$, c_{23} denotes $\cos(\theta_2 + \theta_3)$, $x = \sin 70°$, $y = \cos 70°$.

$$
\begin{aligned}
n_x &= [(c_1 c_{23} c_4 + s_1 s_4)c_5 - y(c_1 c_{23} s_4 - s_1 c_4)s_5 - x c_1 s_{23} s_5]c_6 \\
&\quad - y\left[(c_1 c_{23} c_4 + s_1 s_4)s_5 + y\left(c_1 c_{23} s_4 - s_1 c_4\right)c_5 + x c_1 s_{23} c_5\right] s_6 \\
&\quad - x[x(c_1 c_{23} s_4 - s_1 c_4) - y c_1 s_{23}]s_6
\end{aligned} \tag{2.22}
$$

$$n_y = [(s_1 c_{23} c_4 - c_1 s_4)c_5 - y(s_1 c_{23} s_4 + c_1 c_4)s_5 - x s_1 s_{23} s_5]c_6$$
$$- y\left[(s_1 c_{23} c_4 - c_1 s_4)s_5 + y\left(s_1 c_{23} s_4 + c_1 c_4\right)c_5 + x s_1 s_{23} c_5\right]s_6$$
$$- x[x(s_1 c_{23} s_4 + c_1 c_4) - y s_1 s_{23}]s_6 \tag{2.23}$$

$$n_z = -(s_{23} c_4 c_5 - y s_{23} s_4 s_5 + x c_{23} s_5)c_6$$
$$+ y(s_{23} c_4 s_5 + y s_{23} s_4 c_5 - x c_{23} c_5)s_6$$
$$+ x(x s_{23} s_4 + y c_{23})s_6 \tag{2.24}$$

$$o_x = [-(c_1 c_{23} c_4 + s_1 s_4)c_5 + y(c_1 c_{23} s_4 - s_1 c_4)s_5 + x c_1 s_{23} s_5]s_6$$
$$- y\left[(c_1 c_{23} c_4 + s_1 s_4)s_5 + y\left(c_1 c_{23} s_4 + s_1 c_4\right)c_5 + x c_1 s_{23} c_5\right]c_6$$
$$- x[x(c_1 c_{23} s_4 - s_1 c_4) - y c_1 s_{23}]c_6 \tag{2.25}$$

$$o_y = [-(s_1 c_{23} c_4 - c_1 s_4)c_5 + y(s_1 c_{23} s_4 + c_1 c_4)s_5 + x s_1 s_{23} s_5]s_6$$
$$- y\left[(s_1 c_{23} c_4 - c_1 s_4)s_5 + y\left(s_1 c_{23} s_4 + c_1 c_4\right)c_5 + x s_1 s_{23} c_5\right]c_6$$
$$- x[x(s_1 c_{23} s_4 + c_1 c_4) - y s_1 s_{23}]c_6 \tag{2.26}$$

$$o_z = (s_{23} c_4 c_5 - y s_{23} s_4 s_5 + x c_{23} s_5)s_6$$
$$+ y(s_{23} c_4 s_5 + y s_{23} s_4 c_5 - x c_{23} c_5)c_6$$
$$+ x(x s_{23} s_4 + y c_{23})c_6 \tag{2.27}$$

$$a_x = -x[(c_1 c_{23} c_4 + s_1 s_4)s_5 + y(c_1 c_{23} s_4 - s_1 c_4)c_5 + x c_1 s_{23} c_5]$$
$$+ y[x(c_1 c_{23} s_4 - s_1 c_4) - y c_1 s_{23}] \tag{2.28}$$

$$a_y = -x[(s_1 c_{23} c_4 - c_1 s_4)s_5 + y(s_1 c_{23} s_4 + c_1 c_4)c_5 + x s_1 s_{23} c_5]$$
$$+ y[x(s_1 c_{23} s_4 + c_1 c_4) + y s_1 s_{23}] \tag{2.29}$$

$$a_z = x(s_{23} c_4 s_5 + y s_{23} s_4 c_5 - x c_{23} c_5) - y(x s_{23} s_4 + y c_{23}) \tag{2.30}$$

$$p_x = -x d_6[(c_1 c_{23} c_4 + s_1 s_4)s_5 + y(c_1 c_{23} s_4 - s_1 c_4)c_5 + x c_1 s_{23} c_5]$$
$$+ y d_6[x(c_1 c_{23} s_4 - s_1 c_4) - y c_1 s_{23}] + x d_5(c_1 c_{23} s_4 - s_1 c_4)$$
$$- y d_5 c_1 s_{23} + a_3 c_1 c_{23} - d_4 c_1 s_{23} + a_2 c_1 c_2 + a_1 c_1 \tag{2.31}$$

$$p_y = -x d_6[[s_1 c_{23} c_4 - c_1 s_4]s_5 + y[s_1 c_{23} s_4 + c_1 c_4]c_5 + x s_1 s_{23} c_5]$$
$$+ y d_6[x[s_1 c_{23} s_4 + c_1 c_4] - y s_1 s_{23}] + x d_5(s_1 c_{23} s_4 + c_1 c_4)$$
$$- y d_5 s_1 s_{23} + a_3 s_1 c_{23} - d_4 s_1 s_{23} + a_2 s_1 c_2 + a_1 s_1 \tag{2.32}$$

$$P_z = x d_6(s_{23} c_4 s_5 + y s_{23} s_4 c_5 - x c_{23} c_5) - y d_6(x s_{23} s_4 + y c_{23})$$
$$- x d_5 s_{23} s_4 - y d_5 c_{23} - a_3 s_{23} - d_4 c_{23} - a_2 s_2$$

Eq. (2.18) describes the position of the end coordinate frame $x_6 - z_6$ relative to the base frame $x_0 - z_0$, which is the basis for the kinematics analysis of the robot. If the transformation of the tool coordinate frame $x_t - z_t$ relative to the coordinate frame $x_6 - z_6$ is known, the transformation of the tool coordinate system relative to the base system is

$$_T^0 T = {}_6^0 T \, _T^6 T \tag{2.33}$$

Given a set of joint variables, we can obtain a pose of the end coordinate frame $x_6 - z_6$ according to Eq. (2.21). According to Eq. (2.33), we can get the position of the end point of end effector in the base coordinate frame.

In the robot zero pose, initial values for each joint angle are $\theta_1 = \theta_3 = \theta_5 = \theta_6 = 0, \theta_2 = \theta_4 = -90°$. Corresponding transformation matrices are as follows:

$$_1^0 T = \begin{bmatrix} 1 & 0 & 0 & 0 \\ 0 & 1 & 0 & 0 \\ 0 & 0 & 1 & 0 \\ 0 & 0 & 0 & 1 \end{bmatrix}; \quad _2^1 T = \begin{bmatrix} 0 & 1 & 0 & a_1 \\ 0 & 0 & 1 & 0 \\ 1 & 0 & 0 & 0 \\ 0 & 0 & 0 & 1 \end{bmatrix}; \quad _3^2 T = \begin{bmatrix} 1 & 0 & 0 & a_2 \\ 0 & 1 & 0 & 0 \\ 0 & 0 & 1 & 0 \\ 0 & 0 & 0 & 1 \end{bmatrix};$$

$$_4^3 T = \begin{bmatrix} 0 & 1 & 0 & a_3 \\ 0 & 0 & 1 & d_4 \\ 1 & 0 & 0 & 0 \\ 0 & 0 & 0 & 1 \end{bmatrix}; \quad _5^4 T = \begin{bmatrix} 1 & 0 & 0 & 0 \\ 0 & y & -x & -xd_5 \\ 0 & x & y & yd_5 \\ 0 & 0 & 0 & 1 \end{bmatrix}; \quad _6^5 T = \begin{bmatrix} 1 & 0 & 0 & 0 \\ 0 & y & x & xd_6 \\ 0 & -x & y & yd_6 \\ 0 & 0 & 0 & 1 \end{bmatrix};$$

$$_6^0 T = {}_1^0 T {}_2^1 T {}_3^2 T {}_4^3 T {}_5^4 T {}_6^5 T = \begin{bmatrix} 0 & 0 & 1 & yd_5 + d_4 + a_1 \\ 1 & 0 & 0 & 0 \\ 0 & 1 & 0 & -xd_5 + a_3 + a_2 \\ 0 & 0 & 0 & 1 \end{bmatrix} \tag{2.34}$$

Setting the end point of the sixth joints as P, its coordinates in the coordinate frame $x_6 - z_6$ are $^6P = [0, 0, 0, 1]^T$. Then we determine its coordinates in the coordinate frame $x_0 - z_0$ by matrix transformation:

$$^0 P = {}_6^0 T {}^6 P = \begin{bmatrix} 0 & 0 & 1 & d_6 + yd_5 + d_4 + a_1 \\ 1 & 0 & 0 & 0 \\ 0 & 1 & 0 & -xd_5 + a_3 + a_2 \\ 0 & 0 & 0 & 1 \end{bmatrix} \begin{bmatrix} 0 \\ 0 \\ 0 \\ 1 \end{bmatrix}$$

$$= \begin{bmatrix} d_6 + yd_5 + d_4 + a_1 \\ 0 \\ -xd_5 + a_3 + a_2 \\ 1 \end{bmatrix} \tag{2.35}$$

From the geometrical relationship, it is easy to determine that $^0P_x = d_6 + yd_5 + d_4 + a_1$, $^0P_y = 0$ and $^0P_z = -xd_5 + a_3 + a_2$.

2.1.4 Inverse Kinematics

The inverse kinematics of the robot is to solve the corresponding joint variables given the position and pose of the robot's end coordinate frame. Robot inverse kinematics problem-solving plays an important role in robotics research. It is the basis for studying robot dynamics and robot control. Any progress made will have an important influence on robot's trajectory planning, position control, and off-line programming.

The inverse kinematics solution is much more complicated compared to the forward kinematics. The latter has a uniqueness, i.e. after each joint variable is given, the posture of the end effector or tool is uniquely determined. The inverse kinematics solution is actually the process of solving the kinematics equation. Considering the singularity of the matrix, the inverse of the kinematics equation is uncertain, i.e. it may be unsolved, and there may be only one solution or many solutions. There are many difficulties in solving the direct solution, and it is very difficult to establish a general algorithm.

Figure 2.10 The robot of Motoman HP6.

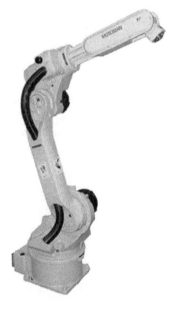

There are many methods for solving the inverse kinematics. Some typical methods are the reciprocal algorithms proposed by Paul et al. [4], the geometric method proposed by Lee and Ziegler [5–7], and the Pieper method [8].

Next, we take the industrial robot, Motoman Hp6 (shown in Figure 2.10) as an example to analyze its inverse kinematics. Its D-H coordinate frames are shown in Figure 2.11. Its D-H parameters are shown in Table 2.3.

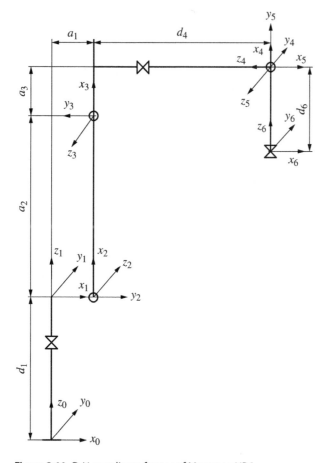

Figure 2.11 D-H coordinate frames of Motoman HP6.

Table 2.3 D-H parameters of Motoman HP6.

Joint i	Varible θ_i/degree (initial value)	a_i/mm	d_i/mm	α_i degree	$\sin\alpha_i$	$\cos\alpha_i$
1	$\theta_1(0)$	150	0	−90	−1	0
2	$\theta_2(-90)$	570	0	180	0	−1
3	$\theta_3(0)$	155	0	−90	−1	0
4	$\theta_4(0)$	0	−640	90	1	0
5	$\theta_5(90)$	0	0	90	1	0
6	$\theta_6(180)$	0	−95	180	0	−1

We can obtain each transformation matrix, as shown below:

$$
{}^{i-1}_iT = \begin{bmatrix} c\theta_i & -s\theta_i & 0 & a_{i-1} \\ s\theta_i c\alpha_{i-1} & c\theta_i c\alpha_{i-1} & -s\alpha_{i-1} & -s\alpha_{i-1}d_i \\ s\theta_i s\alpha_{i-1} & c\theta_i s\alpha_{i-1} & c\alpha_{i-1} & c\alpha_{i-1}d_i \\ 0 & 0 & 0 & 1 \end{bmatrix} \tag{2.36}
$$

where $c\theta_i = \cos\theta_i$, $s\theta_i = \sin\theta_i$, $c\alpha_i = \cos\alpha_i$, and $s\alpha_i = \sin\alpha_i$.
The total transformation matrix is:

$$
{}^0_6T = {}^0_1T \cdot {}^1_2T \cdot {}^2_3T \cdot {}^3_4T \cdot {}^4_5T \cdot {}^5_6T = \begin{bmatrix} r_{11} & r_{12} & r_{13} & p_x \\ r_{21} & r_{22} & r_{23} & p_y \\ r_{31} & r_{32} & r_{33} & p_z \\ 0 & 0 & 0 & 1 \end{bmatrix} \tag{2.37}
$$

where:

$$r_{11} = c_1 c_6 (c_{23} c_4 c_5 + s_{23} s_5) - c_1 s_{23} s_4 s_6$$

$$r_{21} = c_4 c_5 c_6 (s_1 c_{23} + c_1) - s_6 (s_1 c_{23} s_4 - c_1 c_4)$$

$$r_{31} = -c_5 c_6 (s_{23} c_4 + c_{23}) + s_{23} s_4 s_6$$

$$r_{12} = -c_1 s_6 (c_{23} c_4 c_5 + s_{23} s_5) - c_1 s_{23} s_4 s_6$$

$$r_{22} = -c_4 c_5 s_6 (s_1 c_{23} + c_1) - c_6 (s_1 c_{23} s_4 - c_1 c_4)$$

$$r_{32} = s_6 (s_{23} c_4 c_5 - c_{23} s_5) + s_{23} s_4 c_6$$

$$r_{13} = c_1(c_{23}c_5 - c_{23}c_4s_5)$$

$$r_{23} = -s_1c_{23}c_4c_5 - c_1s_4s_5$$

$$r_{33} = s_{23}c_4s_5 + c_{23}c_5$$

$$p_x = d_6c_1c_{23}(c_5 - c_4s_5) + a_3c_1c_{23} + d_4c_1s_{23} + a_2c_1c_2 + a_1c_1$$

$$p_y = -d_6(s_1c_{23}c_4c_5 + c_1s_4s_5) + a_3s_1c_{23} + a_2s_1c_2 + a_1s_1$$

$$p_z = d_6(s_{23}c_4s_5 + c_{23}c_5) + d_4c_{23} - a_3s_{23} - a_2s_2 + d_1$$

where: $c_{23} = c_2c_3 + s_2s_3$, $s_{23} = s_2c_3 - c_2s_3$, $c_i = \cos\theta_i$, $s_i = \sin\theta_i$

The transformation matrix $_6^0T$ of the robot end coordinate system relative to the base coordinate system at one moment is set as:

$$_6^0T = \begin{bmatrix} n_x & o_x & a_x & p_x \\ n_y & o_y & a_y & p_y \\ n_z & o_z & a_z & p_z \\ 0 & 0 & 0 & 1 \end{bmatrix} \tag{2.38}$$

Next, we use the algebra method to show how values of joint variables, θ_1, θ_2, θ_3, θ_4, θ_5, and θ_6. are solved. Because the origins of coordinate frames, $x_4 - z_4$, $x_5 - z_5$, and $x_6 - z_6$, are on the same point, $^1p_{60} = [p_{x1}\ p_{y1}\ p_{z1}\ 1]$ is only decided by θ_1, θ_2, and θ_3, as shown in Figure 2.12. $^0p_{60} = [p_x\ p_y\ p_z\ 1]^T$. $^1p_{60} = [p_{x1}\ p_{y1}\ p_{z1}\ 1] = {}_1^0T^{-1}p_{60}$. From the geometrical relationship in Figure 2.12, the following equations can be deduced.

$$\beta_1 = arc\tan 2(p_{z1}, p_{x1} - a_1);$$

$$\cos\beta_2 = \frac{a_2^2 + (p_{x1} - a_1)^2 + p_{z1}^2 - (a_3^2 + d_4^2)}{2a_2\sqrt{(p_{x1} - a_1)^2 + p_{z1}^2}} \tag{2.39}$$

$$\beta_2 = \arccos\frac{a_2^2 + (p_{x1} - a_1)^2 + p_{z1}^2 - (a_3^2 + d_4^2)}{2a_2\sqrt{(p_{x1} - a_1)^2 + p_{z1}^2}}; \tag{2.40}$$

$$\theta_2 = -(\beta_1 + \beta_2) \tag{2.41}$$

$$\theta_3 = \theta_2 + \arctan 2(p_{z1} + a_2\sin\theta_2, p_{x1} - a_1 - a_2\cos\theta_2)$$
$$- \arctan 2(d_4, a_3) \tag{2.42}$$

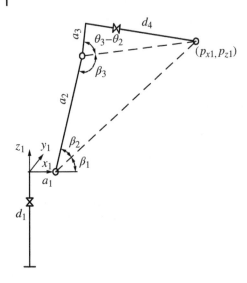

Figure 2.12 Geometrical relationship between first four links in coordinate frame.

From $_6^0T = _1^0T \cdot _2^1T \cdot _3^2T \cdot _4^3T \cdot _5^4T \cdot _6^5T$, we can get the following equation:

$$_6^3T = _3^1T^{-1}\,_1^0T^{-1}\,_6^0T = \begin{bmatrix} _6^3R & ^3P_6 \\ 0 & 1 \end{bmatrix} = _4^3T_5^4T_6^5T \tag{2.43}$$

Now that θ_1, θ_2, and θ_3 are solved, every item of the matrix of $_6^3T$ can be calculated, i.e. they are now constants.

Let

$$_6^3R = \begin{bmatrix} n_{11} & n_{12} & n_{13} \\ n_{21} & n_{22} & n_{23} \\ n_{31} & n_{32} & n_{33} \end{bmatrix}. \tag{2.44}$$

Then: $n_{13} = -\cos\theta_4\sin\theta_5$, $n_{21} = \sin\theta_5\cos\theta_6$, $n_{22} = -\sin\theta_5\sin\theta_6$, $n_{23} = \cos\theta_5$, and $n_{33} = \sin\theta_4\sin\theta_5$.

From these equations, the values of other three joint angles, θ_4, θ_5, and θ_6, can be obtained.

$$\begin{cases} \theta_5 = \arctan 2(-\sqrt{n_{21}^2 + n_{22}^2}, n_{23}) \\ \theta_4 = \arctan 2(-n_{33}, n_{13}) \\ \theta_6 = \arctan 2(n_{22}, -n_{21}) \end{cases} \tag{2.45}$$

When $n_{23} = \cos\theta_5 = 1$, the above solution is degraded. According to the structural features, we make $\theta_4 = 0°$, so:

$$\begin{cases} \theta_5 = 0 \\ \theta_4 = 0 \\ \theta_6 = \arctan 2(-r_{12}, r_{11}) \end{cases} \tag{2.46}$$

2.1.5 Velocity Kinematics

The problem of the robot speed is the problem of the relationship between the joint velocity $\dot{\mathbf{q}}$ of the robot and the speed of its hand movement in the case of the known robot joint position $\mathbf{q}$. The vector of the joint variable is usually expressed as: $\mathbf{q} = \begin{bmatrix} q_1 & q_2 & \cdots & q_n \end{bmatrix}^T$.

The parameter describing the position and attitude of the end effector in the job is called the job coordinate. The task vector, taking the work coordinates as components, usually is written as: $\mathbf{r} = \begin{bmatrix} r_1 & r_2 & \cdots & r_m \end{bmatrix}^T$.

Obviously, the $\mathbf{r}$ is calculated from $\mathbf{q}$:

$$\mathbf{r} = R(\mathbf{q}) \tag{2.47}$$

Generally, $n=m$, and sometimes there are also $n>m$ cases. If we seek the derivative of time on the two sides of the above equation, we can obtain the relationship between the joint workspace speed and the task space speed:

$$\dot{\mathbf{r}} = \frac{dR(\mathbf{q})}{d\mathbf{q}} \dot{\mathbf{q}} \tag{2.48}$$

Let

$$J = \frac{dR(\mathbf{q})}{d\mathbf{q}} = \begin{bmatrix} \dfrac{\partial R_i}{\partial q_j} \end{bmatrix} \tag{2.49}$$

Then

$$\dot{\mathbf{r}} = J\dot{\mathbf{q}} \tag{2.50}$$

J is called the Jacobian matrix, which can be calculated by the homogenous transformation method or vector method [9–11]. When $\dot{\mathbf{r}}$ is given, $\dot{\mathbf{q}} = J^{-1}\dot{\mathbf{r}}$.

2.2 Robot Dynamics

The dynamics problems are similar to the study of kinematic problems, with forward dynamics problems and inverse dynamics problems. Generally speaking, the so-called forward dynamics problem is seeking the movement of the system, given the force acted on the system. The inverse dynamics problem is the opposite: given the movement of the system, find the force acting on the system at this time. Specifically, for the robot, the forward dynamics problem is that given the driving torque τ, find position of joints q, velocities of joints $\dot{q}$, and the robot joint acceleration $\ddot{q}$. For an inverse dynamics problem, for a known set of the robot joint position $\mathbf{q}$, joint speed $\dot{\mathbf{q}}$, and joint acceleration $\ddot{\mathbf{q}}$, find the driving force τ on the robot joints at this time.

The general dynamics equation of a robot is:

$$M(q)\ddot{q} + C(q, \dot{q}) + G(q) = \tau \tag{2.51}$$

where q denotes the vector of joint angles; $M(q)$ is the symmetric, bounded, positive definite inertia matrix, and n is the DoF of the robot arm; $C(q, \dot{q})$ denotes the Coriolis and centrifugal force; $G(q)$ is the gravitational force, and τ is the vector of joint torques. In this equation, the kinetic energy of the robot is described within $M(q)\ddot{q} + C(q, \dot{q})$, and the potential energy is represented in the gravity term $G(q)$. Friction and disturbance input have been neglected here.

There are two commonly used methods for formulating the dynamics in Eq. (2.51), based on the specific geometric and inertial parameters of the robot: the Lagrange–Euler (L–E) formulation and the Recursive Newton–Euler (RN–E) method [12–18]. Both are equivalent, as both describe the dynamic behavior of the robot motion, but are specifically useful for different purposes.

The L–E equations of motion for a conservative system are given by

$$L = K - P, \tau = \frac{\mathrm{d}}{\mathrm{d}t}\left(\frac{\partial L}{\partial \dot{q}}\right) - \frac{\partial L}{\partial q} \tag{2.52}$$

where K and P are the total kinetic and potential energies of the system, respectively, q is the generalized robot coordinates equivalent to joint variables, and τ is the generalized torque at the robot joints. The Newton-Euler equation, it has two parts, the first part describes the translational movement of the mass center,

$$\mathbf{F} + m\mathbf{g} = m\dot{\mathbf{v}}_{\mathrm{c}} \tag{2.53}$$

where **F** is the force acting on the mass center, **g** is the vector of acceleration of gravity, v_c is the linear velocity of the centroid of the rigid body, and m is the total mass. The second part describes the rotational movement of the rigid body,

$$\tau = I\dot{\omega} + \omega \times I\omega \tag{2.54}$$

where τ is the total torque acting on the rigid body, I is the centroidal inertia tensor, and ω is the angular velocity vector.

2.3 Robot Control

2.3.1 Introduction

Robot control is based on robot kinematics and dynamics, and the objective is to make the robot work properly, as expected. For free-moving robots, the purpose is to manipulate the position and posture of the robot's hand. The desired trajectory can be given in the robot's task space and can also be transformed into the desired trajectory in the robot's joint space by inverse kinematics. The desired trajectory is usually in two forms: a fixed position and a trajectory that changes continuously over time. The purpose of the free-moving robot position control is to make the robot hand reach the desired trajectory from any starting position. When the desired trajectory is a fixed-point position, it is called the fixed-point control problem. When the desired trajectory is a continuous trajectory over time, it is called the trajectory-tracking problem. If there are constraints between the robot and the environment, the torques of joints needs to be calculated according to dynamics. In this case, the position of the end-effector is planned, and the forces applying against the environment should also be calculated.

2.3.2 Trajectory Control

In robot control, the position control problem can be described as given a desired trajectory solving joint torques to make the robot manipulator follow that trajectory. Control objectives can be classified into the following two classes [20]: trajectory tracking and regulation.

Trajectory tracking is aimed at following a time-varying joint reference trajectory specified within the manipulator workspace. In general, this desired trajectory is assumed to comply with the actuators' capacity. In practice, the capacities of actuators are set by torque limits, which result in bounds on the acceleration that are complex and state dependent.

Regulation sometimes is also called point-to-point control. A fixed configuration in the joint space is specified. The objective is to bring the joint variable to the desired position and keep it there, in spite of torque disturbances, independent of the initial conditions. The behavior of transients and overshooting are, in general, not guaranteed.

2.3.2.1 Point-to-Point Control

Point-to-point control only controls the poses of the robot hand in some of the specified discrete points in the operating space. The main technical indicators of this control mode are the positioning accuracy and the time required for motion. Point-to-point control is often used in the loading and unloading, handling, spot welding, and plug-in components on the circuit board and other operations in which the positioning accuracy requirement is not high and only the accurate pose of the robot at the target point is required.

Traditionally, control design in robot manipulators can be understood as the simple fact of tuning of a propositional-derivative (PD) or PID compensator at the level of each motor driving the robot joints [19]. Fundamentally, a PD controller is a position and velocity feedback that has good closed-loop properties when applied to a double integrator system. The PID control has a long history since Ziegler and Nichols' PID tuning rules were published in 1942 [20]. Actually, the strong point of PID control lies in its *simplicity* and clear physical meaning. Simple control is preferable to complex control, at least in industry, if the performance enhancement obtained by using complex control is not significant enough. The physical meanings of PID control [21] are as follows:

P control means the present effort making a present state into desired state.

I control means the accumulated effort using the experience information of previous states.

D control means the predictive effort reflecting the information about trends in future states.

PID-type controllers have the advantage of requiring no knowledge of either the model structure or the model parameters.

A simple design method for manipulator control is to utilize a linear control scheme based on the linearization of the system about an operating point. It has the following form:

$$\tau = K_p(q_d - q) - K_v\dot{q} + G(q) \tag{2.55}$$

where K_p and K_v are positive-defined gain matrices and $G(q)$ is the gravitational force. This controller is very useful for set-point regulation, i.e. q_d = constants [22, 23]. When this controller is applied to the general dynamics equation of the robot, the closed-loop equation becomes:

$$M(q)\ddot{q} + C(q,\dot{q}) + K_v\dot{q} - K_pe_q = 0 \tag{2.56}$$

where $e_q = (q_d - q)$. This controller requires knowledge of the gravity components (structure and parameters), though it is simple. K_p and K_v are selected according to the linear control theory to obtain the desired performance index.

An integral action may be added to the previous PD control in order to deal with gravity forces, which to some extent can be considered as a constant disturbance (from the local point of view). The PID regulation controller can be written in the following general form [19]:

$$\tau = K_pe_q + K_i \int fe_q dt - K_v\dot{q} \tag{2.57}$$

where K_i is a positive-definition gain matrix.

2.3.2.2 Trajectories for Paths Specified by Points

While independent PID controls are adequate in most set-point regulation problems, there are many tasks that require effective trajectory tracking capabilities such as plasma-welding, laser-cutting, or high-speed operations in the presence of obstacles. In this case, employing local schemes requires moving slowly through a number of intermediate set points, thus considerably delaying the completion of the task. Therefore, to improve the trajectory tracking performance, controllers should take account of the manipulator dynamic model via a computed-torque-like technique [19].

The tracking control problem in the joint or task space consists of following a given time-varying trajectory qd(t) or xd(t) and its successive derivatives $\dot{q}$d(t) or $\dot{x}$d(t) and $\ddot{q}$d(t) or $\ddot{x}$d(t), which describe the desired velocity and acceleration, respectively. To obtain successful performance, significant effort has been devoted to the development of model-based control strategies [2, 7]. Among the control approaches reported in the literature, typical methods include inverse dynamic control, the feedback linearization technique, and the passivity-based control method.

Inverse Dynamics Control

Inverse dynamics control in joint space has the following form:

$$\tau = M(q)\ddot{q} + C(q, \dot{q}) + G(q) \tag{2.58}$$

Let $v = \ddot{q}$ (v is an auxiliary control input), so:

$$v = \ddot{q}_d + K_V(\dot{q}_d - \dot{q}) + K_P(q_d - q) \tag{2.59}$$

Or, with an integral component

$$v = \ddot{q}_d + K_V(\dot{q}_d - \dot{q}) + K_P(q_d - q) + K_I \int (q_d - q)dt \tag{2.60}$$

Then we get the error dynamics equation

$$\ddot{e}_q + K_V\dot{e}_q + K_P e_q = 0 \tag{2.61}$$

For an auxiliary control input (2.59), or

$$e_q^{(3)} + K_V\ddot{e}_q + K_P\dot{e}_q + K_I e_q = 0 \tag{2.62}$$

Both error dynamics are exponentially stable by a suitable choice of the gain matrices KV and KP (and KI) [19].

Feedback Linearization

The basic idea of feedback linearization is to construct a transformation as a so called *inner-loop control*, which exactly linearizes the nonlinear system after a suitable state space change of coordinates. One can then design a second stage or *outer-loop control* in the new coordinates to satisfy the traditional control design specifications such as tracking, disturbance rejection, etc. [19, 24–27]. The full power of the feedback linearization scheme for manipulator control becomes apparent if one includes in the dynamic description of the manipulator the transmission dynamics, such as the elasticity resulting from shaft windup, gear elasticity, etc. In recent years, an impressive volume of literature has emerged in the area of differential-geometric methods for nonlinear systems. Most of the results in this area are intended to give abstract coordinate -free descriptions of various geometric properties of nonlinear systems and as such are difficult for the nonmathematician to follow [19].

Let $\xi = h(q) + r(t)$, where $h(q)$ is a general predetermined function of the joint coordinate q and $r(t)$ is a general predetermined time function. The control objective will be to select the joint torque inputs τ in order to make the output $\xi(t)$ go to zero [19].

The choice of $h(q)$ and $r(t)$ is based on the control purpose. For example, if $h(q) = -q$ and $r(t) = q_d(t)$, the desired joint space trajectory we would like the manipulator to follow, then $\xi(t) = q_d(t) - q(t) \equiv e_q(t)$ is the joint space tracking error [25]. Forcing $\xi(t)$ to zero in this case would cause the joint variables $q(t)$ to track their desired values $q_d(t)$, resulting in a manipulator trajectory-following problem [19].

To determine a linear state-variable model for manipulator controller design, let us simply differentiate the output $\xi(t)$ twice to obtain [19]

$$\dot{\xi} = \frac{\partial h}{\partial q}\dot{q} + \dot{r} = T\dot{q} + \dot{r} \tag{2.63}$$

$$\ddot{\xi} = T\ddot{q} + \dot{T}\dot{q} + \ddot{r} \tag{2.64}$$

where

$$T(q) = \frac{\partial h(q)}{\partial q} = \left(\frac{\partial h}{\partial q_1} \frac{\partial h}{\partial q_2} \cdots \frac{\partial h}{\partial q_n} \right) \tag{2.65}$$

Given the output $h(q)$, it is straightforward to compute the transformation $T(q)$ associated with $h(q)$.

From $M(q)\ddot{q} + C(q, \dot{q}) + G(q) = \tau$, we can get [19]

$$\ddot{q} = M(q)^{-1}[\tau - \mathbf{n}(\mathbf{q}, \dot{\mathbf{q}})] \tag{2.66}$$

With the nonlinear terms represented by [19]

$$\mathbf{n}(\mathbf{q}, \dot{\mathbf{q}}) = C(q, \dot{q}) + G(q) \tag{2.67}$$

So

$$\ddot{\xi} = T\ddot{q} + \dot{T}\dot{q} + \ddot{r} = \ddot{r} + \dot{T}\dot{q} + T(q)M(q)^{-1}[\tau - \mathbf{n}(\mathbf{q}, \dot{\mathbf{q}})] \tag{2.68}$$

The control input function is defined as:

$$u = \ddot{r} + \dot{T}\dot{q} + T(q)M(q)^{-1}[\tau - \mathbf{n}(\mathbf{q}, \dot{\mathbf{q}})] \tag{2.69}$$

Now we may define a state $y(t)$ by $y = (\xi, \dot{\xi})$ and write the manipulator dynamics as

$$\dot{y} = \begin{bmatrix} 0 & I_P \\ 0 & 0 \end{bmatrix} y + \begin{bmatrix} 0 \\ I_P \end{bmatrix} u \tag{2.70}$$

This is a linear state-space system of the form

$$\dot{y} = Ay + Bu \tag{2.71}$$

driven by the control input u. Due to the special form of A and B, this system is called the *Brunovsky canonical form* and it is always controllable from $u(t)$ [19].

Since [28, 29] is said to be a *linearizing transformation* for the manipulator dynamic equation, one may invert this transformation to obtain the joint torque [19]:

$$\tau = M(q)T^+(u - \ddot{r} - \dot{T}\dot{q}) + \mathbf{n}(\mathbf{q}, \dot{\mathbf{q}})] \tag{2.72}$$

where $T+$ denotes the *Moore–Penrose inverse* of the transformation matrix $T(q)$.

Passivity Based Control Method

This method explicitly uses the passivity properties of the Lagrangian system [30, 31]. In comparison to the inverse dynamics method, passivity-based controllers are expected to have better robust properties because they do not rely on the exact cancellation of the manipulator nonlinearities [19]. The passivity-based control input is given by [19]

$$\dot{q}_r = \dot{q}_d + \alpha e_q, \quad \alpha > 0 \tag{2.73}$$

$$\tau = M(q)\ddot{q}_r + C(q,\dot{q})\dot{q}_r + G(q) + K_V\dot{e}_q + K_P e_q \tag{2.74}$$

The following closed-loop system can be obtained [19]:

$$M(q)\dot{s}_q + C(q,\dot{q})s_q + K_V\dot{e}_q + K_P e_q = 0 \tag{2.75}$$

where $s_q = \dot{e}_q + \alpha e_q$. Construct a Lyapunov function $V(y,t)$ as follows [19]:

$$V(y,t) = \frac{1}{2}y^T \begin{pmatrix} \alpha K_V + K_P + \alpha^2 M & \alpha M \\ \alpha M & M \end{pmatrix} y = \frac{1}{2}y^T P y \tag{2.76}$$

Since the above equation is positive definite, it has a unique equilibrium at the origin, i.e. $y = (e_q^T, \dot{e}_q^T)^T = 0$. Moreover, V can be bounded by

$$\sigma_m \|y\|^2 \le y^T P y \le \sigma_M \|y\|^2, \quad \sigma_M \ge \sigma_m > 0. \tag{2.77}$$

The time derivative of V gives

$$\dot{V} = -\dot{e}_q^T K_V \dot{e}_q - \alpha e_q K_P e_q = -y^T Q y < 0, \tag{2.78}$$

where $Q = \text{diag}[\alpha K_P, K_V]$. since Q is positive definite and quadratic in y, it can also be bounded by

$$k_m \|y\|^2 \le y^T Q y \le k_M \|y\|^2, \quad k_M \ge k_m > 0. \tag{2.79}$$

Then, from the bound of the Lyapunov function V, we get

$$\dot{V} \leq -k_{\mathrm{m}} \, \|y\|^2 = -2\eta V, \quad \eta = \frac{k_{\mathrm{m}}}{\sigma_{\mathrm{M}}} \tag{2.80}$$

This finally yields

$$V(t) \leq V(0) \, e^{-2\eta t} \tag{2.81}$$

It has been shown that the value of α affects the tracking result dramatically [32]. The manipulator tends to vibrate for small values of α. Larger values of α allow better tracking performance and protect s_q from being spoiled by the velocity measurement noise when the position error is small. In [33], it was suggested that [19]

$$K_P = \alpha K_V \tag{2.82}$$

be used for quadratic optimization.

Computed Torque Control

Computed torque control is the technique of applying feedback linearization to nonlinear systems in general [31, 34]. Given the current position and speed of the robot, without considering the viscous friction, and applying the appropriate torque to overcome the inertia of the actuator, the system dynamics equation is:

$$\boldsymbol{\tau} = \mathbf{M}(\mathbf{q})\ddot{\mathbf{q}}_d + \mathbf{C}(\mathbf{q}, \dot{\mathbf{q}}) + \mathbf{G}(\mathbf{q}) \tag{2.83}$$

Now the PD feedback is applied,

$$\ddot{\mathbf{q}} = \ddot{\mathbf{q}}_d + K_V \dot{e}_q + K_P \dot{e}_q \tag{2.84}$$

where $\mathbf{e} = \mathbf{q} - \mathbf{q}_d$. The overall control input becomes

$$\boldsymbol{\tau} = \mathbf{M}(\mathbf{q})(\ddot{\mathbf{q}}_d + K_V \dot{e}_q + K_P \dot{e}_q) + \mathbf{C}(\mathbf{q}, \dot{\mathbf{q}}) + \mathbf{G}(\mathbf{q}) \tag{2.85}$$

And the resulting linear differential equation of the error between the actual trajectory and the desired trajectory are

$$\ddot{e}_q + K_V \dot{e}_q + K_P \dot{e}_q = 0 \tag{2.86}$$

Eq. (2.85) is also called the computed-torque control law, which consists of two parts: the feedback component related to the state error is the compensation torque, which is to eliminate the robot trajectory error; Eq. (2.83) is the feedforward component, which is required torque to drive the system to move along the desired trajectory [35].

2.3.3 Interaction Control

Position control strategies have been successfully used on robotic tasks involving a null or weak interaction between the manipulator and its environment [36]. Common industrial applications are spray painting, welding, and palletising tasks [36–42]. The two main approaches to the control of the interaction of the manipulator and its environment are impedance control and hybrid force-position control [40, 43–46].

2.3.3.1 Impedance Control

The control objective of an impedance controller is to impose, along each direction of the task space, a dynamic relation between the manipulator end effector position and the force of interaction with the environment, the desired impedance. Usually the desired impedance is chosen linear and of second order, as in a mass-spring-damper system [36].

In order to fulfill the task requirements, the user chooses a desired end-effector impedance that may be expressed by [47]:

$$\mathbf{M(q)}\ddot{\mathbf{q}}_d + \mathbf{C(q,\dot{q})} + \mathbf{G(q)} = \mathbf{\tau} \text{-} \mathbf{J}^T F \tag{2.87}$$

where $\mathbf{J}$ is the Jacobian matrix and F is the force between the end effector and the environment.

Let $\dot{X}$ denote the velocity of the end effector in operational space; then $\dot{X} = \mathbf{J}\dot{\mathbf{q}}$ and $a = \ddot{X} = \mathbf{J}\ddot{\mathbf{q}} + \dot{\mathbf{J}}\dot{\mathbf{q}}$.

So

$$\ddot{\mathbf{q}} = \mathbf{J}^{-1}(a - \dot{\mathbf{J}}\dot{\mathbf{q}}) \tag{2.88}$$

Assume X_d is the desired trajectory of the end effector in the operational space and X is the current position of the end effector in the operational space, design a as the PD controller in operational space:

$$a = \ddot{X}_d + K_V(\dot{X}_d - \dot{X}) + K_P(X_d - X) \tag{2.89}$$

From Eqs. (2.87), (2.88), and (2.89), we can obtain:

$$\tau = M(q)(J^{-1}(\ddot{X}_d + K_V(\dot{X}_d - \dot{X}) + K_P(X_d - X) - \dot{J}\dot{q})$$
$$+ \mathbf{C(q,\dot{q})} + \mathbf{G(q)} + J^T F \tag{2.90}$$

2.3.3.2 Hybrid Force-Position Control

There are two extreme states of the robot end effector in contact with the environment. One is that the end effector can move freely in space – that is, there is no contact force between the end effector and the environment. In this case the position control is used. Another is that the end effector and the environment are fixed together, and the end effector cannot change the position, but can apply force and torque in any direction. In this case, the force control is applied. The second case is rare in practice. In most cases, some degrees of freedom are constrained by positional constraints and other degrees of freedom are constrained by force constraints; therefore, it requires the hybrid position-force control [48–50]. The aim of hybrid force/motion control is to split up simultaneous control of both the end-effector motion and contact forces into two separate, decoupled subproblems [19].

In the famous R-C controller [48], the position control law is:

$$\boldsymbol{\tau}_p = \mathbf{M}(\mathbf{q})(\ddot{\mathbf{q}}_d + K_V \dot{e}_q + K_P \dot{e}_q) + \mathbf{C}(\mathbf{q}, \dot{\mathbf{q}}) + \mathbf{G}(\mathbf{q}) \tag{2.91}$$

The force control law is

$$\begin{aligned}\tau_f = M(q)(J^{-1}(\ddot{X}_d + K_V(\dot{X}_d - \dot{X}) + K_P(X_d - X) - \dot{J}\dot{q}) \\ + \mathbf{C}(\mathbf{q}, \dot{\mathbf{q}}) + \mathbf{G}(\mathbf{q}) + J^T F\end{aligned} \tag{2.92}$$

So the hybrid control law is

$$\tau = \tau_p + \tau_f \tag{2.93}$$

Based on this R-C controller, much research has done to improve the performance of hybrid force-motion control [51–56].

References

1 Denavit, J. and Hartenberg, R.S. (1955). A kinematic notation for lower-pair mechanisms based on matrices. *Trans. ASME. J. Appl. Mech.* 22: 215–221.

2 Bajd, T., Mihelj, M., Lenarčič, J. et al. (2010). *Robotics*. Springer.

3 Hu, L. (2012). Structural Scheme Design and Position and Posture errors Analysis for 6-DOF Painting Robot. Dissertation of Huazhong University of Science and Technology.

4 Paul, R.P., Shmano, B.E., and Mayer, G. (1981). Kinematic control equations for simple manipulators. *SMC* 11 (6): 449–455.

5 Lee, C.S.G. and Ziegler, M.A. (1984). Geometric approach in solving the inverse kinematics of PUMA robots. *IEEE Trans. Aerosp. Electron. Syst.* 20 (6): 695–709.

6 Lee, C.S.G. (1982). Robot arm kinematics, dynamics, and control. *Computer* 15 (12): 62–80.

7 Mason, M.T. (2001). *Mechanics of Robotic Manipulation*. Cambridge: MIT Press.

8 Pieper, D. and Roth, B. (1969). The kinematics of manipulators under computer control. In: *Proceedings of the 2nd International Congress on Theory of Machines and Mechanisms*, vol. 2, 159–169.

9 Zong, G.H., Cheng, J.S., and Japan Robot Association (2008). *The New Handbook of Robot Technology*, 2e, 244–245. Beijing: Science Press.

10 Featherstone, R. (1983). Position and velocity transformations between robot end-effector coordinates and joint angles. *Int. J. Rob. Res.* 2 (2): 35–45.

11 Takano, M. (1985). A new effective solution for inverse kinematics problem (synthesis) of a robot with any type of configuration. *J. Fac. Eng. Univ. Tokyo (B)* 38 (2): 107–135.

12 Fu, K. and Gonzalez, C.R.C. (1987). *Robotics Control, Sensing, Vision, and Intelligence*. New York: McGraw-Hill.

13 Bejczy, A.K. (1974). *Robot Arm Dynamics and Control*, 33–669. Pasadena: Jet Propulsion Laboratory Technical Memo.

14 Bejczy, A., Paul, R. (1981). Simplified robot arm dynamics for control. In: IEEE 20th Conference on Decision and Control including the Symposium on Adaptive Processes.

15 Megahed, S.M. (1993). *Principles of Robot Modelling and Simulation*. Hoboken, NJ: Wiley.

16 Lee, C., Lee B. and Nigam R. (1983). Development of the generalized d'alembert equations of motion for mechanical manipulators. In: 22nd IEEE Conference on Decision and Control, pp. 1205–1210.

17 Hollerbach, J.M. (1980). A recursive lagrangian formulation of maniputator dynamics and a comparative study of dynamics formulation complexity. *IEEE Trans. Syst. Man Cybern.* 10 (11): 730–736.

18 Siciliano, B., Sciavicco, L., Villani, L. et al. (2009). *Robotics: Modelling, Planning and Control*. Heidelberg: Springer.

19 Siciliano, B. and Khatib, O. (2008). *Handbook of Robotics*, 135–141. Springer.

20 Ziegler, J.G. and Nichols, N.B. (1942). Optimum settings for automatic controllers. *ASME Trans.* 64: 759–768.

21 Choi, Y. and Chung, W.K. (2004). PID trajectory tracking control for mechanical systems. In: *Lecture Notes in Control and Information Sciences*, vol. 289. New York: Springer.

22 An, C.H., Atkeson, C.G., and Hollerbach, J.M. (1988). *Model-based Control of a Robot Manipulator*. Cambridge: MIT Press.

23 Takegaki, M. and Arimoto, S. (1981). A new feedback method for dynamic control of manipulators. *Trans. ASME J. Dyn. Syst. Meas. Control* 102: 119–125.

24 de Wit, C.C., Siciliano, B., and Bastin, G. (eds.) (1996). *Theory of Robot Control*, 86. Springer.

25 Spong, M.W. and Vidyasagar, M. (1989). *Robot Dynamics and Control*. New York: Wiley.

26 Lewis, F.L., Abdallah, C.T., and Dawson, D.M. (1993). *Control of Robot Manipulators*. New York: Macmillan.

27 Isidori, A. (1985). Nonlinear control systems: an introduction. In: *Lecture Notes in Control and Information Sciences*, vol. 72. New York: Springer.

28 Slotine, J.J. and Li, W. (1991). *Applied Nonlinear Control*. Englewood Cliffs: Prentice-Hall.

29 Rugh, W.J. (1996). *Linear System Theory*, 2e. Upper Saddle River: Prentice-Hall.

30 Berghuis, H. and Nijmeijer, H. (1993). A passivity approach to controller–observer design for robots. *IEEE Trans. Rob. Autom.* 9: 740–754.

31 Slotine, J.J. and Li, W. (1987). On the adaptive control of robot manipulators. *Int. J. Rob. Res.* 6 (3): 49–59.

32 Liu, G. and Goldenberg, A.A. (1996). Comparative study of robust saturation–based control of robot manipulators: analysis and experiments. *Int. J. Rob. Res.* 15 (5): 473–491.

33 Dawson, D.M., Grabbe, M., and Lewis, F.L. (1991). Optimal control of a modified computed–torque controller for a robot manipulator. *Int. J. Rob. Autom.* 6 (3): 161–165.

34 Isidori, A. (1995). *Nonlinear Control Systems*, 3e. New York: Springer.

35 Chen, K., Yang, X., Liu, L. et al. (2006). *Robot Technology and Application*. Tsinghua University Press.

36 Almeida, F., Lopes, A. and Abreu P. (2000). Force-Impedance Control of Robotic Manipulators, Proc. of the 4th Portuguese Conference on Automatic Control, Guimarães, Portugal, 4–6 October.

37 Hogan, N. (1985). Impedance control: an approach to manipulation: part I–III. *ASME J. Dyn. Syst. Meas. Control* 107: 1–24.

38 Hogan, N. (1987). Stable Execution of Contact Tasks Using Impedance Control. Proc. IEEE International Conference on Robotics and Automation, 1047–1054.

39 Mills, J.K. and Goldenberg, A.A. (1989). Force and position control of manipulators during constrained motion tasks. *IEEE Trans. Rob. Autom.* 5: 30–46.

40 Kazerooni, H. (1989). On the robot compliant motion control. *ASME J. Dyn. Syst.Meas. Control* 111: 416–425.

41 De Schutter, J., Bruyninckx, H., Zhu, W. et al. (1997). Force control: a bird's eye view. In: *Control Problems in Robotics and Automation* (ed. B. Siciliano and K.P. Valavanis), 1–18. London: Springer-Verlag.

42 De Schutter, J. and Van Brussel, H. (1988). Compliant robot motion: part I–II. *Int. J. Rob. Res.* 7: 3–33.

43 Kazerooni, H., Sheridan, T.B., and Houpt, P.K. (1986). Robust compliant motion for manipulators: part I–II. *IEEE J. Rob. Autom.* 2: 83–105.

44 Seraji, H. and Colbaugh, R. (1997). Force tracking in impedance control. *Int. J. Rob. Res.* 16: 97–117.

45 Raibert, M. and Craig, J. (1981). Hybrid position/force control of manipulators. *ASME J. Dyn. Syst. Meas. Control* 102: 126–133.

46 Khatib, O. (1987). A unified approach for motion and force control of robot manipulators: the operational space formulation. *IEEE J. Rob. Autom.* 3: 43–53.

47 Shuang, W., Man-lu, L., and Hua, Z. (2017). Simulating study on force tracking and collision avoidance control of robot manipulators. *Autom. Instrum.* 32 (5): 44–49.

48 Raibert, M.H. and Craig, J.J. (1981). Hybrid position/force control of manipulators. *ASME J. Dyn. Syst. Meas. Control* 118: 386–389.

49 Xiong, Y., Tang, L., Ding, H. et al. (1996). *Fundamentals of Robot Technology*. The Press of Huazhong University of Science and Technology.

50 Bassi, E., Benzi, F., Capisani, L.M. et al. (2009). Hybrid Position/fore sliding mode control of a class of robotic manipulators[C]// 48th IEEE Conference on Decision and Control held jointly with 2009 28th Chinese Control Conference. Shanghai, China: Institute of Electrical and Electronics Engineers Inc., 2966–2971.

51 Zhang, H. and Paul, R.P. (1985). Hybrid Control of Robot Manipulators. Proc of the IEEE Int Conf on Robotics and Automation.

52 Miyamura, A. and Kimura, H. (2002). Stability of feedback error learning scheme. *Syst. Control Lett.* 45: 303–316.

53 Dao-xiang, G., Ding-yu, X., and Da-li, C. (2007). Robust adaptive position/force hybrid control for constrained robotic manipulators. *Trans. Syst. Simul.* 19 (2): 348–351.

54 Osy Piuk, R. and Kroger, T.A. (2010). Three-loop model-following control structure: theory and implementation. *Int. J. Control* 83 (1): 97–104.

55 Song, G. and Cai, L. (1998). Robust position/force control of robot manipulators during constrained tasks. *IEEE Proc. Control Theory Appl.* 145 (4): 427–433.

56 Kruger, J. and Surdilovic, D. (2008). Robust control of force-coupled human-robot-interaction in assembly processes. *CIRP Annals-Manuf. Technol.* 57: 41–44.

3

Friction and Contact of Solid Interfaces

3.1 Introduction

Friction is resistant force in moving interface of a dynamical system. To quantify friction, surface roughness that possesses statistical properties should be characterized. Friction is concerned with surface topography, and therefore the determination of the real area of contact and the understanding of the mechanism of mating contacts are critical to characterize friction [1–11]. Particularly, the surface physics can help elaborate the friction in terms of the formation of adhesive junctions by interacting asperities and their breakaway by shearing, whereas continuum mechanics quantify friction by interlock and fracture of asperities [12–14].

To address the problem of real sliding asperity contact is quite difficult, which involves in complex dynamics. The feasible approach is to assume the contact to be of a quasi-static nature to certain extent. In many applications with relatively smooth surfaces the deformation of contacting asperities can be assumed to be linear and elastic. For many problems, the contact has to be extended to inelastic and nonlinear conditions and to involve in dynamics [15–25].

In this chapter we present the fundamentals of contact and friction between two contact surfaces in the context of quasi-static state by assuming that the normal motion is ignored. We focus on the mechanics of contact and friction by outlining the mechanical attributes of various friction processes in the context of the problems of the friction-dynamics of artificial arm.

After the introduction, the second section briefs the mathematical description of surface roughness and introduces the fundamental contact mechanics including Hertz analytical solution of single asperity

Dynamics and Control of Robotic Manipulators with Contact and Friction, First Edition.
Shiping Liu and Gang (Sheng) Chen.
© 2019 John Wiley & Sons Ltd. Published 2019 by John Wiley & Sons Ltd.

contact and the treatment of multiple asperity contact between two solid surfaces. The third section presents the basic principles of friction of two contact surfaces. The basic physics of adhesion of two surfaces are presented. The friction of two dry contact surfaces is presented, and then the friction of liquid mediated interfaces is described. In the last section, varied friction models are presented.

3.2 Contact Between Two Solid Surfaces

3.2.1 Description of Surfaces

In nature and in engineering practice, there are no topographically smooth surfaces. All solid surfaces, however, smooth in a given scale, are comparatively rough in a smaller scale. In other words, solid surfaces are almost always comprised of random variations in surface height in some scale as shown in Figure 3.1. This feature of surface repeats down to nano-scale. A surface that looks like smooth visually is actually quite rough under microscale investigation. These surface height deviations are called as surface roughness that is characterized

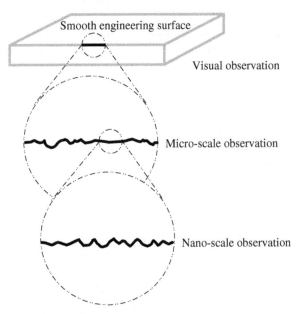

Figure 3.1 Roughness in different scales.

by asperities and valleys of various magnitudes. Surface roughness is associated with a length scale. The magnitude of the surface height deviations can be different at different length scales for the same surface. Usually, the surface fluctuations with a long wavelength are called as the waviness of a surface, whereas those having a shorter wavelength are commonly associated with the roughness.

Surface roughness is usually quantified by use statistics. Statistic parameters like the distribution of the surface heights are used to describe the vertical deviations of surface height with respect to a reference plane. Most surface roughness is random and the distribution of the surface heights usually follows a Gaussian distribution. The most common measures of surface roughness include the centerline-average Ra characterizing the average roughness and root mean square roughness Rq. We consider a surface height profile $z(x)$ over a length L in which profile heights is measured from a mean line. The centerline-average is defined as

$$R_a = \frac{1}{L} \int_0^L |z - m| dx \tag{3.1}$$

in which the mean is given by

$$m = \frac{1}{L} \int_0^L z \, dx \tag{3.2}$$

The root mean square (rms) roughness is defined as

$$R_q = \sqrt{\frac{1}{L} \int_0^L z^2 dx} \tag{3.3}$$

The variation of the distribution is characterized by the standard deviation of the height of the surface from the center-line, σ, which is defined as

$$\sigma = \sqrt{\frac{1}{L} \int_0^L (z - m)^2 dx} = \sqrt{R_q^2 - m^2} \tag{3.4}$$

Skewness (dimensionless) of roughness is

$$\gamma = \frac{1}{L\sigma^3} \int_0^L (z - m)^3 dx \tag{3.5}$$

Kurtosis (dimensionless) of roughness is

$$\kappa = \frac{1}{L\sigma^4} \int_0^L (z - m)^4 dx \tag{3.6}$$

A more detailed statistics analysis follows. If we denote probability density function by $p(z)$ to be the probability that the height of a particular point in the surface will lie between z and $z + dz$, then the probability that the height of a point on the surface is greater than z_0 is given by the cumulative probability function:

$$P(z \leq z_0) = \int_{-\infty}^{z_0} p(z)dz = P(z_0) \tag{3.7}$$

It has been found that many real surfaces exhibit a height distribution that is close to the following normal or Gaussian probability function,

$$p(z) = \frac{1}{\sigma\sqrt{2\pi}} e^{\left[-\frac{(z-m)^2}{2\sigma^2}\right]} \tag{3.8}$$

3.2.2 Contact Mechanics of Two Solid Surfaces

When two elastic spheres are brought together in contact and then loaded, local deformation at contact regime will occur. The contact point will enlarge into a surface of contact. Its area is a circle when the contacting bodies are spheres. The contact area size and pressure between them can be determined from their geometry and elastic properties. The solutions for deformation, area of contact, pressure distribution, and stresses at the contact area were quantified by conventional Hertz theory. The related assumption is made that the deformation is elastic, and the dimensions of the contact area are small, relative to the radii of curvature and to the overall dimensions of the bodies. Thus, the radii, though varying, may be taken as constant over the very small areas surrounding the contact area. The deflection integral is going to be derived for a plane surface.

Assume two contact spheres of different material and radii R_1 and R_2, respectively. Figure 3.2 shows the spheres after loading, with the radius a of the contact area greatly exaggerated for illustration.

Assume the effective modulus involving Young's moduli and Poisson ratios, E_1, E_2, v_1, and v_2 is defined as follows:

$$E^* = \left[\frac{(1 - v_1^2)}{E_1} + \frac{(1 - v_2^2)}{E_2}\right]^{-1} \tag{3.9}$$

This expression describes the compliance of the system at a fixed contact area. Furthermore, Hertz theory gives the relationship between load and deformation of the two contact spheres as the two

Figure 3.2 Contact of two spheres.

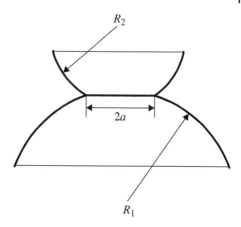

surfaces are brought into contact under load. It had been derived that the contact radius a is given by

$$a = \left(\frac{3RP}{4E^*} \right)^{1/3} \tag{3.10}$$

in which P is load. Here R is a function of R_1 and R_2, called the radii of curvature:

$$R = \left[\frac{1}{R_1} + \frac{1}{R_2} \right]^{-1}$$

Define the contact displacement as

$$\delta = R_1 + R_2 - d \tag{3.11}$$

where d is the distance between the centers of the two spheres. The compressive displacement or the normal approach is given by

$$\delta = \frac{a^2}{R} \tag{3.12}$$

the contact area is given by

$$A = \pi a^2 = \pi R \delta \tag{3.13}$$

The formula can be used for many other cases, such as one surface being concave. When one surface is concave, the contact area is on the inside of a surface. For this situation, the numerical value of its radius is to be taken as negative in all equations.

In the above theory, the assumption is that the two surfaces in contact are frictionless, so that shear stresses are assumed to be not developed at the interface.

This "frictionless" assumption is often appropriate for very stiff materials where adhesive forces are relatively unimportant, but it is often not the case for softer materials where adhesive forces play a very important role. In these cases, a "full-friction" boundary condition, where sliding of the two surfaces is not allowed, is often more appropriate. Hertz model is applicable to the nonadhesive cases of the load and displacement. Actual values of the load and displacement will generally differ from these values when the contacting surfaces adhere to one another. There are numerous studies [26–80] dedicated to extending the application of Hertz theory, some of which will be presented in the subsequent section.

All engineering surfaces are not smooth to a certain extent. The asperities on the surface of very compliant solids could be squashed flat by the contact pressure, and perfect contact could be likely obtained through the contact area. It is unusual to flatten rough surfaces by elastic or plastic deformation of the asperities as shown in the Figure 3.3.

In general, contact between solid surfaces is discontinuous, and the real area of contact is a small fraction of the nominal contact area. The real contact surfaces are actually the asperity contact summation and are usually much smaller than the apparent surface. Next, we discuss the contact of identical asperities and the random asperities.

Although in general all surfaces have roughness, certain simplification can be achieved if the contact between a single rough surface and a perfectly smooth surface is considered.

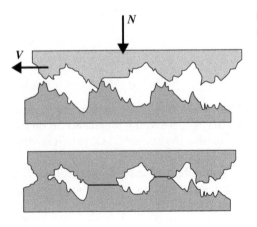

Figure 3.3 Contact surfaces.

Moreover, the problem will be simplified further by introducing a theoretical model for the rough surface in which the asperities are considered as sphere array so that their elastic deformation characteristics may be defined by the Hertz theory. It is further assumed that there is no interaction between separate asperities, that is, the displacement due to a load on one asperity does not affect the heights of the neighboring asperities.

Figure 3.4 shows a surface of unit nominal area consisting of an array of identical spherical asperities all of the same height z with respect to some reference plane xx'. The number of asperity is N. As the smooth surface approaches the rough surface, due to the application of a load, the normal approach is given by $(z - d)$, where d is the current separation between the smooth surface and the reference plane.

Clearly, each asperity is deformed equally and carries the same load P_i so that for N asperities per unit area the total load P will be equal to NP_i. For each asperity, the load P_i and the area of contact A are known from the Hertz theory. Thus if R is the asperity radius, then

$$P_i = \frac{4E^*}{3} R^{1/2} \delta^{3/2} \tag{3.14}$$

and $A_i = \pi R \delta$. The total load is

$$P = \frac{4E^* N A_i^{3/2}}{3\pi^{3/2} R} \tag{3.15}$$

The load can be related to the total contact area $A = NA_i$ by

$$P = \frac{4E^* A^{3/2}}{3\pi^{3/2} N^{1/2} R} \tag{3.16}$$

This result indicates that the real area of contact is related to the two-thirds power of the load, when the deformation is elastic.

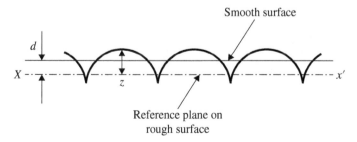

Figure 3.4 Contact of sphere array against with smooth surface.

It has been discussed that the asperities of real surfaces have different heights characterized by a probability distribution of their peak heights. Therefore, the simple surface models should be modified accordingly and the analysis of its contact must now include probability estimation as to the number of the asperities in contact. If the separation between the smooth surface and the reference plane xx' is d, and then there will be a contact at any asperity whose height was originally greater than d as shown in Figure 3.5.

The conventional statistical model dealing with contact surfaces with asperities is Greenwood and Williamson model, which assumes that the summit of asperities on surface is spherical, the asperity summits on two surfaces have a constant radius, R_{p1} and R_{p2} respectively, and a composite radius defined below is used to characterize interface.

$$R_p = \left[\frac{1}{R_{p1}} + \frac{1}{R_{p2}} \right]^{-1},\tag{3.17}$$

The height is assumed to be random and with Gaussian distribution, with standard deviations σ_{p1}, σ_{p2} and an equivalent standard deviation is defined as follows,

$$\sigma_p = (\sigma_{p1}^2 + \sigma_{p2}^2)^{1/2}\tag{3.18}$$

For elastic contact in static condition or in dynamic condition without shear stress, GW model quantifies the relationship apparent pressure p_a, apparent contact area A_a, mean real pressure p_r, real contact area A_r, number of contact number n as a function of separation d. Then the problem reduces to a plate in contact with asperity. If $p(z)$ is the probability density of the asperity peak height distribution, then the probability that a particular asperity has a height between z and

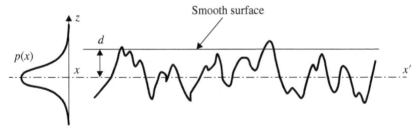

Figure 3.5 Rough surface.

$z + dz$ above the reference plane will be $P(z)$. Thus, the probability of contact for any asperity of height z is

$$P(z > d) = \int_d^\infty p(z)dz \tag{3.19}$$

If we consider a unit nominal area of the surface containing asperities, assume the total asperity number is N, then the number of contacts n will be given by

$$n = N \int_d^\infty p(z)dz \tag{3.20}$$

Since the normal approach is $(z - d)$ for any asperity, the total area of contact and the expected load will be given by

$$A_r = \pi N R_p \int_d^\infty (z - d)p(z)dz \tag{3.21}$$

$$P = A_r p_r = A_a p_a = \frac{4}{3}NE^* R_p^{1/2} \int_d^\infty (z - d)^{3/2}p(z)dz \tag{3.22}$$

It is convenient to use non-dimensional variables. According to [4, 24], the derived relationship can be written as,

$$\frac{p_a}{(\eta R_p \sigma_p)E^*(\sigma_p/R_p)^{1/2}} = \frac{4}{3}F_{3/2}(D) \tag{3.23}$$

$$\frac{p_r}{E^*(\sigma_p/R_p)^{1/2}} = \frac{4}{3\pi}F_{3/2}(D)/F_1(D) \tag{3.24}$$

$$\frac{A_r E^*(\sigma_p/R_p)^{1/2}}{p_a A_a} = \frac{3\pi}{4}F_1(D)/F_{3/2}(D) \tag{3.25}$$

$$(A_r/n)R_p\sigma_p = \pi F_1(D)/F_0(D) \tag{3.26}$$

$$\frac{nR_p\sigma_p E^*(\sigma_p/R_p)^{1/2}}{p_a A_a} = 3F_0(D)/4F_{3/2}(D) \tag{3.27}$$

Using these equations one may evaluate the total real area, load, and number of contact spots for any given height distribution, in which $D = d/\sigma_p$ is the dimensionless separation, $\eta = N/A_a$ is the density of asperity summits per unit area on a surface with smaller density, and $F_m(D)$ is a function defined as

$$F_m(D) = \int_D^\infty (s - D)^m p^*(s)ds \tag{3.28}$$

where $p^*(s)$ is the standardized peak-height probability density function in which the height distribution has been scaled to make its standard deviation unit. A usually used approximation is given by,

$$A_r \approx \frac{3.2 p_a A_a}{E^*(\sigma_p/R_p)^{1/2}} \qquad (3.29)$$

From which the real area of contact is proportional to the load. The relationship between the real area of contact and the load will be dependent on both the mode of deformation and the distribution of the surface profile. More supplicated analysis is based on fractal roughness description and continuum plasticity theory. When the asperities deform elastically, depending on the deformation mode within the contact, its real area can be estimated from:

$$A_r = c\left(\frac{P}{E^*}\right)^n \qquad (3.30)$$

in which $2/3 < n < 1$, the linearity between the load and the real area of contact occurs only where the distribution approaches an exponential form and this is very often true for many practical engineering surfaces.

When two surfaces of high elastic modulus materials are pressed together, they only come into contact at the tips of the asperities. The total contact area is A_r and $A_r \ll A_a$, the nominal area of contact. For this reason, the stress on the asperities is generally large. When the stress exceeds yield pressure, the plastic deformation occurs.

When analyzing the real contact between two engineering surfaces, it is usually assumed that they are covered with asperities having random height distribution and deforming elastically or plastically under normal load. The sum of all micro-contacts created by individual asperities constitutes the real area of contact which is usually only a tiny fraction of the apparent geometrical area of contact. There are two groups of properties, namely, deformation properties of the materials in contact and surface topography characteristics, which define the magnitude of the real contact area under a given normal load P. Generally, the contact behavior of solid in contact is determined by a value defined as plasticity index, ψ,

$$\psi = \left(\frac{E^*}{H}\right)\left(\frac{\sigma_p}{R_p}\right)^{1/2} \qquad (3.31)$$

In which E^* is the composite or effective elastic modulus, H is hardness, σ_p and R_p are the composite standard deviation and composite radius of summits (surface heights).

If the plasticity index $\psi < 0.6$, then the deformation is largely elastic. In the case when $\psi > 1.0$, the predominant deformation mode within the contact zone is plastic deformation. The index depends on both the mechanical properties and the surface roughness of the contact surfaces.

The mechanical property ratio E^*/H and the surface roughness determine the extent of plasticity in the contact region. For most metals and ceramics, $E^*/H > 100$. For polymer it is on the order of 10. Thus the plasticity index for polymer is on the order of one-tenth of that of metal and ceramics, therefore, the contact is primarily elastic except for very rough surfaces. On the other hand, for surface with very small surface roughness the contact is usually elastic.

For plastic contact, we have the following approximation,

$$A_r = c\frac{P}{H} \tag{3.32}$$

where c is the proportionality constant. When the asperities deform plastically, the load is linearly related to the real area of contact for any distribution of asperity heights.

The introduction of an additional tangential load produces a phenomenon called junction growth which is responsible for a significant increase in the asperity contact areas. The magnitude of the junction growth of metallic contact can be estimated from the expression

$$A = A_{p0}\left[1 + \alpha\left(\frac{F_f}{P}\right)^2\right]^{1/2} \tag{3.33}$$

In which A_{p0} is the real area of contact without any shear stress, α is a constant, F_f is friction force, P is normal load.

3.3 Friction Between Two Solid Surfaces

3.3.1 Adhesion

When bringing two solid surfaces into contact, adhesion, or bonding across the interface develops. This leads to the adhesive force that is perpendicular to the surface, in addition to the applied normal force.

Adhesion could occur both in solid–solid contact and liquid involved solid–solid contact.

Two clean solid surfaces tend to create strong bond, whereas contaminated or boundary film covered surfaces tend to yield weak bonds.

The adhesive junctions of asperities on solid–solid contact are caused by inter-atomic and intermolecular force attractions. Generally, the adhesion could be chemical or physical. The former includes covalent bonds, ionic or electrostatic bond, metallic bonds, and hydrogen bonds, the physical interaction involves the van der Waals bonds.

The adhesion is generally proportional to the normal force. This is because the applied normal force increases the real contact area that promotes the bonds. On the other hand, the real contact area also increases as a result of inter-atomic attraction. Moreover, an additional shear force in addition to the normal force is usually increase the adhesive force, as the shear effect in addition to the compressive effect tends to increase the real contact area.

There are many references on the solid–solid adhesion friction [16, 81–113], some of recent developments incorporate the results of numerical analysis and experimental investigation for the elastic–plastic contact, adhesion, and sliding asperity in modern statistical representation of roughness. In the following we describe some of the basic theories, which cover interfacial forces such as electrostatic forces, van der Waals Forces.

Electrostatic bonding occurs when positively charged cations and negatively charged anions have interaction. Electrostatic attractive forces across the interface can arise from a difference in work functions or from electrostatic charging of opposed surfaces. Difference in the work function leads to the formation of an electrical double layer by a net transfer of electrons from one surface to the other.

At small separations the electrostatic pressure between flat surfaces is generally lower than the van der Waals pressure. It can be assumed that all objects are free of charge at the beginning before contact. However, after contacting the objects, contact electrification and triboelectrification occur, and forces due to these charges could occur.

The free energy for electrostatic interaction between two charged atoms or ions with a distance of x is given by

$$E_s = \frac{z_1 z_2 e^2}{4\pi\varepsilon_0\varepsilon x} \tag{3.34}$$

the electrostatic force is obtained by differentiating the energy with respect to distance x as,

$$F_s = -\frac{z_1 z_2 e^2}{4\pi\varepsilon_0\varepsilon x^2} \tag{3.35}$$

in which e is the charge of a single electron, $e = 1.6 \times 10^{-19}$ C, z_1 and z_2 are the ionic valences, ε_0 is the permittivity of a vacuum 8.854×10^{-12} $C^2 \, Nm^{-2}$, ε is the dielectric constant of the medium, 1.000 59 in air at 1 atm.

Consider electrostatic force between two parallel plates with surface charge density of σ on one of the plates, the electric field E is given by

$$E = \frac{\sigma}{\varepsilon_0 \varepsilon} \qquad (3.36)$$

Then the electrostatic force per unit area is given by

$$F_s = -\frac{1}{2} \varepsilon \varepsilon_0 E^2. \qquad (3.37)$$

If the potential difference between the plates is V, then electric field is $E = V/x$.

van der Waals force always occurs between molecules and are much smaller than the binding force between atoms. The van der Waals force is effective from comparatively large separation (up to 10 nm) down to interatomic spacing (about 0.2 nm). The van der Waals dispersion forces between two bodies are caused by mutual electric interaction of the induced dipoles in the two bodies. Dispersion forces generally dominate over orientation and induction forces except for strongly polar molecules. van der Waals forces exist for every material in every environmental condition and depends on the object geometry, material type, and separation distance.

When two atoms are brought close enough, they start to experience the intermolecular forces interacting each other. At the beginning, it is an attraction force, and its strength increases with decreasing distance until a maximum point is reached, then it decreases with decreasing distance. When the distance is reduced further, the force becomes repulsive and increasingly stronger. This is reflected in Figure 3.6, which shows the potential energy between two atoms as a function of their distance. One usually expression used to describe this potential is the Lennard-Jones potential, in which the attractive van der Waals potential is modeled as an inverse sixth power term and the repulsive potential is modeled as an inverse twelfth power term,

$$W = -\frac{C}{r^6} + \frac{D}{r^{12}}, \qquad (3.38)$$

in which $C = 10^{-77} \, Jm^6$ and $D = 10^{-134} \, Jm^{12}$ are two constants for atoms in vacuum.

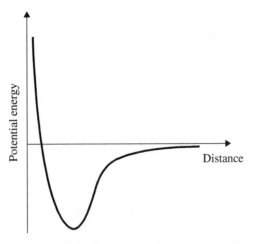

Figure 3.6 Potential energy of molecules.

For interface with a few nano-meters spacing, the repulsive potential term can be ignored, and we obtain the purely van der Waals potential,

$$W = -\frac{C}{r^6} \tag{3.39}$$

Equation (3.38) is the potential between two atoms, and it can be integrated over an infinitely long and infinitely deep half space to obtain the potential between an atom and an infinite plate,

$$W = -\frac{\pi C \rho_1}{6h^6} + \frac{\pi D \rho_1}{45h^9} \tag{3.40}$$

where ρ_1 is the number density of atoms in the infinite plate, and h is the distance between the atom and the plate. Equation (3.40) can be further integrated over a volume of material to get the potential between an amount of material and an infinite plate.

To obtain the intermolecular force between an amount of material with surface S and the plate, we need to differentiate the integrated potential in the direction perpendicular to the plate. After that, the intermolecular force between each S of material and the plate can be written as

$$F_v = \frac{dW_v}{dz} = \frac{A}{6\pi} \iint_S \frac{dxdy}{h^3} - \frac{B}{45\pi} \iint_S \frac{dxdy}{h^9} \tag{3.41}$$

where A is the Hamaker constant, B is the constant related to the repulsive term, The first term on the right-hand side of Eq. (3.41) is the attractive van der Waals force, and the second term is the repulsive intermolecular force. The attractive and repulsive portions of the force have different acting ranges. The attractive van der Waals force has a much longer acting range than the repulsive portion.

Based on Eq. (3.41), the intermolecular force per unit area can be simplified as

$$p_v = \frac{A}{6\pi h^3} - \frac{B}{45\pi h^9} \tag{3.42}$$

In which A is Hamaker constant, B is the repulsive term related constant.

The two solid surfaces first experience the attraction force when the distance between them is less than about 10 nm. The strength of the attractive force increases with the reduction of the spacing until $h_a = (2B/5A)^{1/6}$ reaching maximum attractive value. When the spacing is further reduced, the short range repulsive force becomes effective and in the end makes intermolecular force be zero at $h_0 = (2B/15A)^{1/6}$. Below this threshold value the repulsive force will be dominant, then other interactions like the re-arrangement of surface could happen due to rapidly increase of repulsive force.

As an approximation to solids, assume $A = 10^{-19}$ J, and $B = 10^{-76}$ Jm^6 we get $h_a \approx 0.3$ nm. An approximation of the interaction energy due to van der Waals forces per unit area between two parallel plates in the non-retarded regime ($h < 20$ nm) can be given by the attractive term only,

$$E_v = -\frac{A}{12\pi h^2}$$

and the intermolecular force per unit area between the plates can be approximated by,

$$p_v = -\frac{A}{6\pi h^3} \tag{3.43}$$

in which A is Hamaker constant. For most solids and liquids, the Hamaker constant lies in the range 0.4–4×10^{-19} J.

An approximation to precise calculation of van der Waals force is to apply the concept of free surface energy [3, 4, 7]. Assume two materials have free surface energies per unit area of γ_1, γ_2, the energy of adhesion per unit area is defined as

$$W_a = \Delta\gamma = \gamma_1 + \gamma_2 - \gamma_{12} \tag{3.44}$$

in which $\Delta\gamma$ equals to a reduction in the surface energy of the interface per unit area. It is negative and it represents the energy that needs to be applied to separate a unit bonded interface. Furthermore, it can be approximated $\Delta\gamma = C(\gamma_1 + \gamma_2)$, in which C is the compatibility parameter for the two materials, and always fall in between 1 and 0.

In most cases, a kind of assumption can be made for micro asperities, but an exact solution would be more proper for the nano asperities.

The intermolecular attractive forces depend on atomic spacing and the corresponding surface energies of materials. For super smooth surface, the adhesion can be very larger leading to the virtual welding of one surface to another. For rough or particles involved interface, its effects can be neglected.

For sufficiently small size contacts, the normal adhesion forces between the surfaces affect the contact conditions. Various adhesion models, between an elastic sphere and a flat have been introduced to extend the Hertz model. The model by Johnson, Kendall, and Roberts (JKR) [27] assumes that the attractive intermolecular surface forces cause elastic deformation beyond that predicted by the Hertz theory and produce a subsequent increase of the contact area. This model also assumes that the attractive forces are confined to the contact area and are zero outside the contact area.

When the surface adhesion force is considered, the stress distribution between two surfaces is tensile at the edge of the contact area and remains compressive in the center. JKR model is based on these considerations.

Consider the energy release rate, w, which has the units of a surface energy (energy/area), and describes the amount of energy that is needed to decrease the contact area, A, by a unit amount,

$$w = \frac{\partial}{\partial A}(U_E + U_M) \tag{3.45}$$

where U_E is the elastic energy of the system and U_M is the mechanical potential energy associated with the applied load. In terms of the current values of P, R, and a, the contact radius was found to be,

$$a = \left[\frac{3R}{4E^*}(P + 3\pi Rw + \sqrt{6\pi RPw + (3\pi Rw)^2})\right]^{\frac{1}{3}} \tag{3.46}$$

where w is the energy per unit contact area, equal to the thermodynamic work of adhesion and E^* is the equivalent elastic modulus of the two spheres. As a result of the surface forces, the contact size is larger than the value in the Hertz model and will be finite for zero external force.

The contact force P_1 between the surfaces is bigger than normal load P:

$$P_1 = P + 3\pi Rw + \sqrt{6\pi RwP + (3\pi Rw)^2}. \tag{3.47}$$

The corresponding contact displacement is

$$\delta = \frac{(3P_1)^{\frac{2}{3}}}{(4E^*)^{\frac{2}{3}} R^{\frac{1}{3}}} - \frac{1}{2E^*} \left(\frac{4E^*}{3RP_1} \right)^{\frac{1}{3}} (P_1 - P) \tag{3.48}$$

which depends nonlinearly on the external force P. The JKR model is based on the analysis at equilibrium conditions and has been verified by static and quasi-static experiments.

Another model of this type assumes that the contact area does not change due to the attractive surface forces and remains the same as in the Hertz theory; thereby the attractive forces are assumed to act only outside of the contact area. More models were developed to fill the gap between the above two models.

Liquid involved interface has extra adhesion in addition to the above discussed adhesions acting on dry solids [114–158]. Liquid-mediated adhesion generally consists of meniscus force and viscous force.

Viscous force is rate-dependent and is only significant for high viscous liquid. Meniscus force is due to surface tension. The presence of the capillary condensation and liquid film could significantly increase the adhesion between surfaces.

Particularly, even for a dry interface, moisture (or other liquid vapor) could condense from vapor onto the interface as bulky liquid, and exhibit annular-shaped capillary condense around the contact asperities. The schematic of liquid condensation in interface is shown in Figure 3.7. This phenomenon is very common in engineering interfaces, particularly for having exposure in ambient environment where water moisture is unavoidable.

A thin liquid layer between two solid plates can work as an adhesive. If the contact angle between liquid and solid is less than 90 as shown in Figure 3.8, the pressure inside the liquid will be lower than outside and a net attractive force between the plates will appear.

Assume θ is the contact angle between liquid and solid in air. h is liquid thickness, and A is wet area. F is external force. The pressure difference Δp_{la} at the liquid–air interface is given by the Laplace equation,

$$\Delta p_{la} = \frac{\gamma_{la}}{r} \tag{3.49}$$

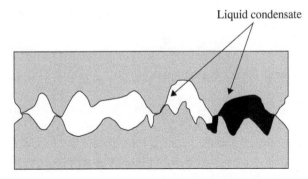

Figure 3.7 Liquid condensation in interface.

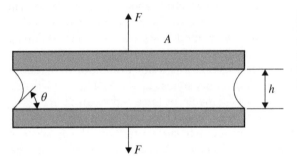

Figure 3.8 A thin layer liquid working as an adhesive between two plates.

where γ_{la} is the surface tension of the liquid–air interface, and r is the radius of curvature of the meniscus (negative if concave).

Consider the liquid is between the plates and the liquid contacts the solid at the fixed contact angle. From simple geometry it follows that

$$r = -\frac{h}{2\cos\theta} \tag{3.50}$$

In equilibrium, an external force F separating the plates must be applied to counterbalance the capillary pressure forces:

$$F = -\Delta p_{la}A = \frac{2A\gamma_{la}\cos\theta}{h} \tag{3.51}$$

where A is the wetted area. Note that a positive force F corresponds to a negative Laplace pressure. The pressure inside the liquid is lower than outside and the plates are pushed together by pressure forces.

This equation can also be used to evaluate the meniscus between a solid sphere contacting a plate with liquid meniscus.

For stiction calculations it is convenient to calculate the surface energy stored at the interface that is bridged by a drop of liquid. Consider a drop of liquid placed on a solid surrounded by air, in equilibrium, the contact angle between liquid and solid is determined by the balance between the surface tensions of the three interfaces. This balance is expressed by Young's equation:

$$\gamma_{sa} = \gamma_{sl} + \gamma_{la} \cos\theta, 0 < \theta < \pi \tag{3.52}$$

where γ_{sa} is the surface tension of the solid–air interface and γ_{sl} is the surface tension of the solid–liquid interface. Young's equation is also valid for configurations other than the typical one. The contact angle is the same on a curved or irregular shaped surface.

Consider a general case of solid sphere proximity to a surface with separation of D forming a meniscus. The surface has continuous liquid film h as shown in Figure 3.9, the meniscus is given by,

$$F = -2\pi R\gamma(1 + \cos\theta) \tag{3.53}$$

When surface has no liquid film, or $h = 0$, the meniscus is given by

$$F = -4\pi R\gamma \cos\theta / (1 + D/(S - D)) \tag{3.54}$$

In real interface, stiction can show a large dependence on the relative humidity of air. Friction measurements of silicon and ceramics show a strong dependence of the static friction coefficient on relative humidity. This is caused by the meniscus force due to capillary condensation. Both the normal and horizontal components of meniscus force could contribute to the friction. Liquids that wet or have a small contact angle on surfaces will spontaneously condense into cracks, pores, and into small gaps surrounding the points of contact between the contacting surfaces. At equilibrium the meniscus curvature is characterized by Kelvin radius:

$$r_K = \frac{V\gamma_{la}}{RT \ln(p/p_s)} \tag{3.55}$$

Figure 3.9 Schematic of an isolated meniscus in the presence of a liquid film.

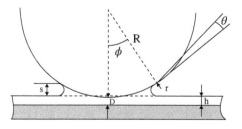

where V is the molar volume ($1.804 \times 10^{-5} m^3/mol$ at $20\,°C$), p is the vapor pressure and p_s is the saturation vapor pressure. At room temperature, $V\gamma_{la}/RT = 0.54$ nm for water. The meniscus curvature strongly depends on the relative vapor pressure p/p_s. At 100% relative humidity, a water film can grow all over the surface. The amount of condensed liquid in thermodynamic equilibrium is determined by both the Kelvin radius and the contact angle. The meniscus curvatures are equal to the Kelvin radius and the contact angles satisfy Young's equation.

Models for the adhesion force due to capillary condensed liquid have been fully developed in hard disk drive tribology.

Kinetic meniscus force can be estimated as follows [4]. When a surface approaches another one with the liquid mediation, there is kinetic process leading to the final equilibrium. During this process, the flow of liquid could increase the meniscus. The driving pressure can be represented as

$$p = -\frac{\gamma_{la}}{r} - \frac{A}{6\pi h^3} \tag{3.56}$$

The first term is due to Laplace equation, the second terms originates from van der Waals effect.

Assume the volume of liquid to be v, its flow rate is,

$$\frac{dv}{dt} = -\frac{2\pi r h^3}{3\eta_l} \frac{\partial}{\partial r} \left[\frac{\gamma_{la}}{r} + \frac{A}{6\pi h^3} \right] \tag{3.57}$$

then the radius of projected region of meniscus, neck radius $x(t)$, can be derived as a function of time with respect to its equilibrium value $(x)_{eq}$, finally, the meniscus force can be given by

$$F(t) = 2\pi R\gamma (1 + \cos\theta) \left[\frac{x(t)}{(x)_{eq}} \right]^2 \tag{3.58}$$

A drop of lubricant filling in the gap between the two surfaces will form a toe-dipping regime and thus initiate the stiction. With the increase of the contact time, in addition to the solid–solid asperity micro-displacement, the micro-flow and the diffusion of the lubricant will redistribute the interface lubricant to form an equilibrium state, and therefore form pillbox regime accompanying the micro-descent process. This is schematically shown in Figure 3.10. Many efforts have been reported to quantify the meniscus force.

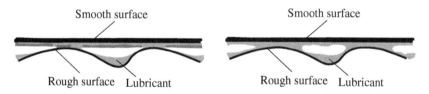

Figure 3.10 Schematic of rough asperities on a disk in contact with a slider surface: (a) Short-term: toe-dipping regime; and (b) long-term: pillbox regime.

Another meniscus model is given in the following. As shown in Figure 3.9, the hemispherical surface of an asperity with radius R on a solid is in near contact with (distance D) the flat surface of the other solid. The lubricant build-up is with a transient thickness h on the solid flat surface.

The meniscus force can be derived as

$$F_m = 2\pi R^2 \gamma (1 + \cos\theta)\Phi \frac{d\Phi}{dD} \qquad (3.59)$$

where γ is the surface tension of the lubricant, Θ is the contact angle formed by the lubricant; D is the distance between the flat plane and the hemispherical surface of the asperity.

In addition to the meniscus force, the viscous force also contributes to the adhesive force by the following relationship:

$$F_v = \frac{\beta \eta_l}{t_s} \qquad (3.60)$$

in which β is a proportional constant, η_l is the dynamic viscosity of the liquid, and t_s is the time to separate the two surfaces. Obviously, t_s is inversely proportional to the breakaway acceleration and velocity of interface.

3.3.2 Dry Friction

3.3.2.1 Friction Mechanisms

When there is contact sliding between two bodies under a normal load, P, a resistant force always exists – this force is called frictional force, F. The frictional force acting on a component always acts in a direction opposite to the motion of the component. Two basic facts about friction have been experimentally established: (i) the frictional force is a function of the normal load on the contact, $F = \mu P$, where μ is the coefficient of friction and P is the normal load; (ii) the frictional force is independent of a nominal area of contact.

These two statements constitute what are known as the laws of sliding friction under dry conditions [159–192].

Studies of sliding friction have a long history. But there is still no simple model that could be used by an engineer to calculate the frictional force for a given pair of materials in contact. It is now widely accepted that friction results from complex interactions between contacting bodies. This includes the adhesion at the points of contact due to the molecular interaction, the effects of surface asperity deformation, as well as the plastic deformation of a weaker material. Figure 3.11 shows the schematic of normal contact and slope contact of asperities, the former tends to develop adhesions, the latter can develop both adhesion and asperity deformation in horizontal direction. A number of factors, such as the mechanical and physical–chemical properties of the materials in contact, surface topography, operational, and environment conditions as well as system dynamics determine the dominant components in friction process. At a fundamental level there are several major factors that determine the friction of dry solids:

i) real area of contact;
ii) shear strength of the adhesive junctions formed at the points of real contact;
iii) breakaway of these junctions during relative motion;
iv) system dynamics.

The friction properties of a given material/component are not its intrinsic properties, they depend on many factors related to a specific

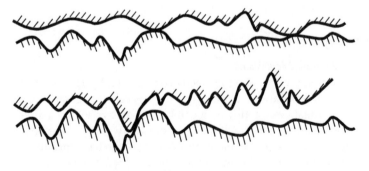

Figure 3.11 The normal contact and the slope contact of asperities.

applications. Quantitative values for friction in the forms of friction coefficient depend on the following basic groups of parameters:

i) The friction pair, i.e. the components and their relevant properties of mechanical, physical, chemical aspects, and geometry;
ii) The interface, i.e. the formation and breakaway of contact junctions in multiple scales;
   The operational variables, i.e. load (pressure) and speed and time;
   The environmental variables, i.e. temperature and humidity;
iii) The system dynamics, the mutual interaction of the system's components and their time dependent variables (note that the interface is time-dependent variable due to wear and layer forming).

Moreover, coefficient of friction is also complicated by other factors, like the scales. For instance, for a given interface, coefficient of friction could be inversely proportional to the normal load, however, when load down to micro- to nano- Newton level, the coefficient of friction could have different feature.

Friction is always associated with energy dissipation and conversion.

Mechanical energy is transformed within the real area of contact, mainly through elastic deformation and hysteresis, plastic deformation, plow, and adhesion.

Dissipation of mechanical energy takes place mainly through: thermal dissipation (heat), storage within the bulk of the body by the generation of defects, cracks, and emission like acoustic emission and thermal generation.

For typical contact interface, the friction is given by

$$F = A[\alpha \tau_a + (1 - \alpha)\tau_l] \tag{3.61}$$

in which τ_a, τ_l are respectively the average shear stress of the dry contact and of the liquid film. α is the fraction of dry area. The average shear stress of the liquid film is contributed by both viscosity of the liquid and meniscus effect,

$$\tau_l = \frac{\eta_l V}{h} + \tau_m \tag{3.62}$$

in which η_l is the dynamic viscosity of the liquid, V is the relative sliding velocity, and h is the liquid film thickness, τ_m is the shear stress due to meniscus effect. In boundary lubrication conditions or the interfaces exposed to humid environments, the meniscus, and even viscous effect could be significant.

In the next we discuss dry friction due to adhesion. We have shown that the area of real contact between two bodies is usually a very small fraction of the apparent geometrical area of contact. However, the friction force is determined almost entirely by the area of real contact.

Consider two dry surfaces, the upper body moves parallel to the lower body with the velocity v. The friction force is given by,

$$F = \tau_a A_r \tag{3.63}$$

where A_r is the real contact area and τ_a is shear stress.

One of the most important components of friction originates from the formation and breakaway of interfacial adhesive bonds. Extensive theoretical and experimental studies have explained the nature of adhesive interaction. The main emphasis was on the electronic structure of the bodies in frictional contact. From a theoretical point of view, attractive forces within the contact zone include all those forces which contribute to the cohesive strength of a solid, such as the metallic, covalent, and ionic short-range forces as well as the secondary van der Waals bonds which are classified as long-range forces. The interfacial adhesion is as natural as the cohesion which determines the bulk strength of materials.

The coefficient of friction due to adhesion can be approximated as the ratio of the interfacial shear strength of the adhesive junctions to the yield strength of the asperity material,

$$\mu_a = \frac{F_a}{P} = \frac{A_r \tau_a}{P} \approx \frac{\tau_a}{p_r} \tag{3.64}$$

in which p_r is the mean real pressure. Substitute A_r from Eq. (3.29), we obtain,

$$\mu_a \approx \frac{3.2\tau_a}{E^*(\sigma_p/R_p)^{1/2}} \tag{3.65}$$

For single asperity case, it can be approximated as,

$$\mu_a \propto \frac{1}{P^{1/3}} \tag{3.66}$$

For plastic contact,

$$\mu_a \approx \frac{\tau_a}{H} \tag{3.67}$$

In the elastic contact case like a diamond like coat coated disk against ceramic slider, μ_a decreases with an increase in roughness. In the plastic contact case, μ_a could be independent with roughness in moderate range. It can be seen that μ_a tends to high at very smooth surface due to the growth of real area of contact, and also tends to high at very rough surface due to interlocking.

Evaluation of μ_a needs to quantify τ_a, which can be estimated by a limit analysis. Generally, the shear strength τ_a cannot substantially exceed the bulk shear strength for plastic contacts.

For most engineering materials this ratio τ_a/p_r is of the order of 0.2 and therefore the dry friction coefficient could be of the same order of magnitude. In the case of clean metals, where the junction growth is most likely to take place, the adhesion component of friction may increase to about 10–100 times.

The presence of any type of surface oxide layer, lubricant, or water film preventing the formation of the adhesive junction can dramatically reduce the magnitude of the adhesion component of friction.

The real area of contact could be much larger than that is from the deformation due to applied load, this is because of the work of adhesion. Assume the attack angle or the slope of the asperity is θ, the modification of Eq. (3.64) with adhesion work can be derived by letting the work done by the normal force be equal to the work done in deformation and the change in the surface energy. Using this equilibrium, the above simple model can be supplemented by the surface energy of the contacting components, the friction coefficient is given by

$$\mu_a = \frac{\tau_a}{p_r}\left[1 - \frac{2w\tan\theta}{p_r}\right]^{-1} \tag{3.68}$$

where w is the surface energy.

Next we discuss the plastic effect. Suppose the material has a yield stress σ_y. If the initial area of contact is small, the asperities in contact will yield and deform plastically in compression. The area of contact will grow as these asperities are squashed and others come into contact. Eventually the normal force $P = A_r\sigma_y$, then the true area of contact will be

$$A_r = P/\sigma_y \tag{3.69}$$

When the sliding asperities past each other, the frictional force Fs relates to average shear stress $\tau = F_s/A_r$ acting on the asperities. If the contact at the asperities is intimate and the asperities have deformed extensively, the asperities may have to shear plastically in order to slide. In this case, $\tau = \tau_a$, in which τ_a is the yield stress in shear. Since the yield stress in tension and compression is $\sigma_y = 2\tau_a$ in terms of Tresca criterion,

$$\tau_a = \sigma_y/2, \tag{3.70}$$

therefore

$$F_s = A_r\sigma_y/2 \tag{3.71}$$

consider $A_r = P/\sigma_y$, we have $F_s = 0.5P$. The approximation of the friction law for asperity deformation yields the friction coefficient being 0.5. This is close to the value found for dry solids including metals, ceramics, and glasses.

More typically, solids like metals oxidize to some extent in air and form a thin oxide layer on their surfaces. The interface between two oxide surfaces may slide at stresses less than τ_a of the metal, leading therefore to a $\mu_a < 0.5$.

Note that the formation of strong junction bond needs acting time for asperity creep deformation. Once sliding occurs, the alternative asperity attach/detach and the oscillations in normal direction make the time be insufficient to achieve full atom-to-atom bonding over the entire contact area, therefore the coefficient of friction could decrease in sliding process. This is one of the reasons that kinetic friction is smaller than dynamic friction.

Actual values of cof may range from as high as 5 to as low as 0.05, depending on the materials brought into contact and the conditions of their surfaces. For pure metals in contact the local contact areas, "junctions," between asperities grow as sliding occurs due to extensive plastic deformation around the asperities. This growth in junction area leads to extensive bonding and a cof > 5. This phenomenon is part of the basis of "friction welding" of metals and plastics. If surfaces are rapidly rubbed past each other, the heat generated can melt low melting materials such as the resin in bowed string.

For metals with clean surfaces, the interfacial shear force can be approximated by [108],

$$\tau_s = c_0 + c_1 p + c_2 p^2 \tag{3.72}$$

In which c_0, c_1, c_2 are material constants. p is normal pressure.

The above discussion applies to sliding contacts between the metals, where A_r is small but σ_y and τ_a are large so that cof is large. For other kind of sliding pairs, the situation depends on the real contact area A_r, σ_y, and τ_a.

For ceramics on ceramics, e.g. slider-disk interface in hard disk drives, the hardness and σ_y is high, the contact area is small and the tendency to bond across the contact is small and τ_a is small and thus cof is small.

Rigid polymer films on harder materials like Teflon (PTFE) on metals yield small coefficient of friction, this is because the polymer chains are oriented along the direction of sliding and only weak van der Waals bonds between chains need to be broken.

In addition to the adhesion, elastic and plastic deformation in continuum interfaces, and fracture process also helps develop the friction. Consider the mode of failure due to crack propagation, the friction due to the fracture of an adhesion junction was derived,

$$F = \frac{c\sigma_{12}\delta_c}{n^2(PH)^{1/2}} \tag{3.73}$$

in which c is coefficient, σ_{12} is the interfacial tensile strength, P is normal load, δ_c is the critical crack opening displacement, n is the work-hardening factor, H is the hardness.

In the next we discuss the dry friction due to deformation (slope). Besides the microscope interface interaction of smooth surface where asperity normal contact dominated, the micro and macroscopic interaction of asperity of rough surfaces could occurs and consequently asperity deformation and plow of soft material by hard material usually occur.

In this sliding process, the mechanical energy is dissipated through the deformations of contacting bodies. The slip-line (lines of maximum shear stress) model of continuum mechanics could be used to analyze the deformation of the single surface asperity. Figure 3.12 shows a slip-line field-based deformation model of friction based on Prandtl two-dimensional stress analysis. The material in the region ABCDE flows downward and outward as the hard material moves

Figure 3.12
Schematic of slip-line.

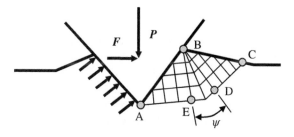

forward. Three distinct regions of plastically deformed material may develop in these regimes. The flow shear stress of the material defines the maximum shear stress which can be developed in these regions [5, 15]. The angle ψ can be chosen so that the velocities of elements of material on the free surface, contact surface, and boundary of the rigid plastic material are consistent. The coefficient of friction is given by,

$$\mu_d = \lambda \tan \left\{ \arcsin \left[\frac{\sqrt{2}(2 + \psi)}{4(1 + \psi)} \right] \right\} \tag{3.74}$$

where λ is the portion of plastically supported load, and it is a function of elastic modulus and hardness.

The proportion of load supported by the plastically deformed regions is a complicated function of the ratio of the hardness to the elastic modulus. For completely plastic asperity contact and an asperity slope of 45°, the coefficient of friction is 1.0. It decreases to 0.55 for an asperity slope approaching zero.

Another approach to this problem is to assume that the frictional work performed is equal to the work of the plastic deformation during steady state sliding. This energy-based plastic deformation model of friction gives the following expression,

$$\mu_d = \frac{A_r}{P} \tau_m \left\{ 1 - 2 \frac{\ln\left(1 + \frac{\tau_s}{\tau_m}\right) - \left(\frac{\tau_s}{\tau_m}\right)}{\ln\left[1 - \left(\frac{\tau_s}{\tau_m}\right)^2\right]} \right\} \tag{3.75}$$

where A_r is the real area of contact, τ_m denotes the ultimate shear strength of a material, and τ_s is the average interfacial shear strength.

The problem of relating friction to surface topography in most cases reduces to the determination of the real area of contact and studying the shear stress.

If one surface is harder than the other, the asperities of hard one could penetrate into the soft one. It leads to groove if the shear strength is exceeded. Plowing not only increase friction, it also creates wear particle which change friction as well. To maintain the plowing motion, a force is required and it may constitute a major component of the overall frictional force. The schematic of a plowing of soft material by the hard, conical asperity is shown in Figure 3.13a. Assume the effective

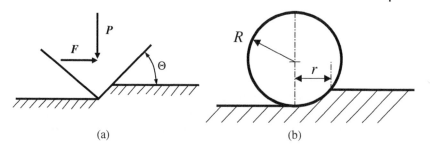

Figure 3.13 Plowings of hard conical and sphere asperity against soft elastic substrates.

slope is Θ, the formula for estimating the coefficient of friction in this case is as follows,

$$\mu_p = \frac{2}{\pi} \tan \Theta \tag{3.76}$$

For most engineering materials, the slope angle of asperity (roughness angle) are very small (5–10°), and the plowing component of friction is correspondingly very small. A typical case is that effective slope is less than 10°, and the coefficient μ_p is about 0.05 and less. In elastic contact, μ_pa) is often assumed to be negligibly small.

Frictional force due to plowing is very sensitive to the ratio of the radius of curvature of the asperity to the depth of penetration. Consider a spherical asperity of radius R in contact with a softer body as shown in Figure 3.13b, r is the size of the plowing area that is proportional to the depth of penetration. The coefficient of friction is given by,

$$\mu_p = \frac{2}{\pi} \left\{ \left(\frac{R}{r}\right)^2 \sin^{-1}\left(\frac{r}{R}\right) - \left[\left(\frac{R}{r}\right)^2 - 1\right]^{1/2} \right\} \tag{3.77}$$

The cof increases rapidly with an increase of r/R, and cof increases as sphere digs deeper.

Moreover, plowing brittle materials is associated with microcracking process, a model developed in terms of fracture mechanics to quantify the effects of fracture toughness, elastic modulus, and hardness on cof is given as,

$$\mu = c\frac{K_k^2}{E(HP)^{1/2}} \tag{3.78}$$

In which K_K is the fracture toughness. E is the elastic modulus and H is the hardness.

For rough hard interface or hard interface involved particles, the plowing could be the dominant factor for energy dissipation. Plastic materials could undergo plastic deformation. Viscoelastic material could exhibit hysteresis process. During the sliding, the softer material is pressed and then the stress is released as moving forward and contact move on. Each time the contact portion is stressed and energy is stored, when progress to next contact portion, most of the energy released as the stress is removed. Only small portion is dissipated as heat when hysteresis losses.

Friction is a complex functional which is dependent on a large variety of parameters, such as sliding speed, temperature, normal load, humidity, surface topography, and surface films. To further elaborate friction mechanism, we consider energy dissipation during friction. In engineering interfaces, several friction mechanisms could co-exist and interact with each other in a complicated way.

In general, frictional work is dissipated at two different locations within the contact zone.

The first location is the interfacial region characterized by high rates of energy dissipation and usually associated with an adhesion mechanism of friction. The other one involves the bulk of the body and the larger volume of the material subjected to deformations where the rates of energy dissipation are much lower. Energy dissipation during plowing and asperity deformations takes place in this second location. The above distinction of two locations is artificial and only for the purpose of simplification of a very complex problem. Actually, in describing KJL modeling we illustrate the basic energy balance equation which indicates the interaction aspect from contact point of view. The various processes can be briefly characterized by one or several items as follows:

i) plastic deformations and micro-cutting;
ii) viscoelastic deformations leading to fatigue cracking and tearing, and subsequently to subsurface excessive heating and damage;
iii) true sliding at the interface leading to excessive heating and thus creating the conditions favorable for chemical degradation;
iv) interfacial shear creating transferred films;

During sliding contact, part of the kinetic energy produces waves and oscillations in the bodies, and part of it leads to plastic and elastic deformation of asperity tips. Some energy expends through viscous dissipation, and the balance through adhesion, fracture, chemical reactions, and photoemission. Distribution of energy conversion through this process varies for different applications. Each of these processes provides a mechanism for converting the original kinetic energy to an interim one in the form of vibration and sound, deformation energy, surface energy, tribo-chemical energy, and other tribo-emissions. In the end, part of the initial energy remains stored as potential energy, and part of it converts to thermal energy, eventually dissipating to the surroundings. Thus, friction can be viewed as a combination of processes that transforms ordered kinetic energy into either potential energy or disorderedly thermalized state of kinetic energy. It then follows that the friction force can be considered as a combination of forces that resist motion during each of these energy conversion processes,

$$
F_f = F_{adhesion} + F_{elastic\ deformation} + F_{plastic\ deformation}
$$
$$
+ F_{viscousity} + F_{fracture} + F_{random} \tag{3.79}
$$

The schematic depiction illustrates the contributions to friction at different scales.

To characterize the above process affecting friction needs different models from macro to micro or even smaller scale, as the various event take place in each scales.

For the integration of the contribution from these scales, requires expertise from different disciplines of science and engineering. Usually, a continuum scale is suitable for modeling engineering problems. At the atomic scale, the primary problems relate to the dissipation and oscillation of atoms will be examined in later section.

Many of the current efforts to model friction start at the continuum scale, relating surface roughness to friction. The obvious mechanisms that contribute to friction in such models include elastic/plastic contacts, viscous dissipation, fracture, and adhesion. Each of the process develops at each true contact region between the surfaces. And, the true contacts take place between asperities on the surfaces or on particles between them.

3.3.2.2 Friction Transitions and Wear

The above we discussed the interface with clean surfaces. For the real engineering surfaces, they are all unlikely to be clean. The oxide films could form on the surface of metal and other engineering materials, yielding a boundary layer. Its thickness could vary from a few atomic thickness to a fraction of a micron. The boundary layers usually have lower shear strength than the original solid. For many engineering surface, the moisture or other chemical vapor could adsorbed on the surface to form adsorbed layer.

The transfer film forms more readily on roughened surface and that the transfer film can exist in solid state and in a low viscosity or fluid state. The formed film in interface have substantial effect on friction.

Friction is always associated wear. They always co-exist and both are the result of one solid rubbing against another and interacts each other. Wear is the effect of friction on material surfaces that rub together. Wear occurs as a result of interaction between two contacting surfaces, and it can be classified as the type of fatigue, adhesion, chemical wear, corrosion, and abrasion. Wear is usually associated with the loss of material from contracting bodies in relative motion. The formation and removal of interface films due to wear could lead to substantial change and variation in friction. Like friction, wear is controlled by the properties of the material, the environmental and operating conditions and the geometry of the contacting bodies. Adhesive wear is invariably associated with the formation of adhesive junctions at the interface. Archard's classical wear model states that the rate of wear, $\dot{V}$, is proportional to the pressure, p, which is given by,

$$\dot{V} = k \frac{pL}{3H} \tag{3.80}$$

where k is the wear coefficient, and L is the sliding distance and H is the hardness of the softer material in contact. Two materials exhibiting the same friction coefficient can exhibit quite different wear rates because the energy is partitioned differently between and within the materials.

Friction is intimately related to adhesion, interlock and wear which forms highly non-equilibrium processes occurring at multiple-levels from nano, molecular level to micro- and macro. It leads what happens at the macroscopic level. The interface can be smooth or rough, hard or soft, elastic, viscoelastic or plastic, brittle or ductile, dry or wet (including lubricated), and of very different physics, chemistries, and thermal dynamics.

Next we consider the energy balance associated with friction and wear process. The energy that is transformed as a consequence of frictional contact can be stored in the friction system or dissipated in a number of different ways. If E_f is the energy resulting from sliding contact, E_{out} is the energy leaving the friction system, and E_{st} is the energy remaining in the friction system, then we have

$$E_f = E_{st} + E_{out} \tag{3.81}$$

For example, mechanical energy from sliding can be converted to heat, vibrations, and sound, to material deformation, or the creation of new surfaces by fracture. Likewise, it can be stored in the material as the energy associated with micro-structural defects, etc.

The typical engineering surfaces are always covered by adsorbed film that consists of water, oxygen, or even oil. For metal surface, there usually have oxide layer. Sliding surfaces are always progressive during sliding process, due to the friction and wear process. The changes in the surface also impact the friction. For interface, after a certain period of sliding, the friction could level off to a steady state. It is called as "break-in," or "run-in" period. After the break-in, the cof could increase to a higher plateau or a lower plateau, exhibiting a s-shaped curve or Z-shaped curve. The break-in process is associated with the peel off of high asperities, surface polish to be more matchable, original film breakage and wear off and new stable film forming. For some material under some limit operation, the severe wear could onsets, and this could give rise to a salient rise in friction.

Another associated problem is the temperature. When two surfaces slide over each other, the heat Q produced is proportional to the sliding velocity and friction force

$$Q = \int_{A_r} V \tau dA_r \tag{3.82}$$

In which Ar is the actual area of contact and τ is the shear stress and V is velocity. Q depends on area of contact Ar, the shear stress and velocity, or indirectly depends on normal load and coefficient of friction. The temperature rise of surface depends on the thermo-mechanical properties of the two contacting bodies as well as Q.

The surface temperature generated in contact areas has a major influence on wear, material properties, and material degradation. The friction process converts mechanical energy primarily into thermal energy which results in a temperature rise.

The friction intensity may not be sufficiently large to cause a substantial temperature rise on the body, but it could be sufficiently large to cause a substantial temperature rise on the surface. The flash temperature is defined as the temperature rise in the contact area above the bulk temperature of the solid as a result of friction energy dissipation. The surface temperature rise can influence local surface geometry through thermal expansion, causing high spots on the surface which concentrate the load and lead to severe local wear. The temperature level can lead to physical and chemical changes in the surface layers. These changes can lead to substantial transitions in friction mechanisms and wear phenomena. Consider a simple formulation for the mean flash temperature in a circular area of real contact of diameter $2a$. The friction energy is assumed to be uniformly distributed over the contact region. For stationary heat source, consider one of the sliding components, component one, the mean temperature increase above the bulk solid temperature is

$$\Delta T = \frac{Q}{4ak_1} \tag{3.83}$$

where Q is the rate of frictional heat supplied to component 1, (Nm s^{-1}), k_1 is the thermal conductivity of body 1, $(\text{W m}^{-1}\,^{\circ}\text{C})$, and a is the radius of the circular contact area (m).

3.3.2.3 Static Friction, Hysteresis, Time, and Displacement Dependence

In the next, we discuss static friction. For contact surfaces, the break force required to initiate motion could be lower, higher, or even equal to the force needed to maintain the surface sliding in the subsequent relative motion. Therefore, the coefficient of static friction could be smaller, higher, or equal to the coefficient of kinetic friction. The higher static friction usually appears for clean surface. For these cases, the static friction is usually a function of dwell time. The static friction increase with time for dry interface is considered due to the interface bonding from the interaction of the atoms on the mating surface and the plastic flow and creep of interface asperities under load. The time-dependent static friction model usually can be represented as

$$\mu(t) = \mu_{\infty} - (\mu_{\infty} - \mu_0)\exp(-\alpha t_s), \tag{3.84}$$

where μ_0 is the initial value of the coefficient of friction. μ_{∞} is the limit value of static friction at long term time, is rest time, α is constant. Other models based on power law has also been developed [86].

In the next we present the hysteresis property of friction occurring in some systems. In acceleration and deceleration processes of certain system, friction vs. velocity curves may not be identical. There usually exists some delay to form hysteresis loop as shown in Figure 3.14. The friction exhibits the memory feature. The state variable models have been developed to quantify this type of characteristics, which has critical applications [87, 88]. In solid mechanics, the constitutive relations have been used effective at capturing steady state and certain transient effects in a wide variety of materials with interfaces having micron scale roughness. The approach involves expressing the friction in terms of the instantaneous slip speed at the interface and one or more state variables, for which phenomenological evolution equations are also introduced.

One of the underlying assumptions is that the interfacial area is large enough to be self-averaging, so that a mean-field-like state variable is sufficient to capture the collective dependence of the microscopic degrees of freedom on the dynamical variables, including time, displacement, slip speed, which characterize the motion.

The state variable models contain a displacement-dependent feature. It not only adopts a steady-state velocity-dependent property incorporating Stribeck curve which will be illustrated in next section, it but also assumes an instantaneous velocity-dependent reflecting the fact that the instant change in velocity results in instantaneous change in friction. Moreover, it assumes an evolutionary characteristic of sliding distance-dependent, which reflects the phenomena that following a sudden change in velocity, a steady-state friction curve is approached through an exponential decay over characteristic sliding distances.

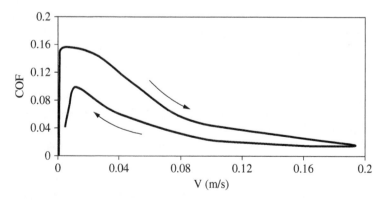

Figure 3.14 Hysteresis of friction-velocity curve.

Various phenomenological models developed offer such kind of features where the coefficients of model are fitted to experiment data and then used to describe a wide range of observed frictional behaviors, such as the dilation of a liquid under shear and the transition between stick–slip (regular or chaotic) and smooth sliding friction. However, most of "state variables" in models are unable to be quantitatively related to the properties of real physical system.

In many applications, the displacement-dependent static friction needs to be characterized. The static friction is actually a constraint and it is associated with an elastic or plastic deformation under traction. The displacement before overcoming static friction can be modeled with an equivalent spring, k, which is a function of asperity, normal force, material elasticity, and surface energy. The pre-sliding displacement has long been studied in many engineering problems.

3.3.2.4 Effects of Environmental and Operational Condition on Friction

First of all, we discuss the temperature-dependent feature of friction. The friction of some interface is highly humidity-dependent. In a humid environment, the amount of condensate water present at the interface increase with the increase of their relative humidity. The adsorbed water film thickness on diamond like carbon-coated solid surface can be approximated as follows,

$$h = h_1(RH) + h_2 e^{\alpha(RH-1)} \tag{3.85}$$

RH is the relative humidity ranging from 0 to 1. h_1, h_2, α are constants. For instance, some existing experiments shows that the coefficient of friction of lubricated solid surface increase rapidly when relative humidity is above a critical value 60%, this critical humidity is dependent upon the interface roughness. Above the critical limit, the static friction increase could be up to five times, whereas the kinetic friction could only increase slightly.

Load and velocity have strong effects on friction. The oxide films are likely produced on the surface of metal and alloys. This usually gives rise to lower friction than clean surface under low load. When load increases, the plowing could occur and film could get penetrated and higher friction could be given.

When a load is very high, the friction gets decreased due to the larger quantity of wear debris and the surface roughing. The coefficient of friction is usually decreases with an increase in velocity. High speed motion tends to reduce the formation of asperity junction of interface. High speed tends to generate heating softening of asperity and the oxidation of film. On the other hand, softening could increase plowing, high shear rates leads to lower real contact area, this renders the friction could be higher in high speed. The real velocity dependence of friction depends on many competing factors. However, for wet interface, coefficient of friction is usually decreases with an increase in velocity when velocity is under certain threshold which will be discussed in next section.

The load and speed dependence of the friction of ceramics could be related to the changes in the chemical surface films and the extent of fracture. The cof is low at low load, but increase rapidly after the brittle fracture occurred. A decrease in friction with sliding velocity is attributed to the increase in the interface temperature which promotes the formation of tribo-chemical films.

In many cases the friction-velocity peak occurs because of the competing effects of strain rate dependence of the shear strength and the reduced contact area due to viscoelastic stiffening of some materials.

3.3.3 Liquid Mediated Friction

3.3.3.1 Stribeck Curve
In the above we discussed the interface with dry friction surfaces or surface friction without liquid involved. For many real engineering interfaces, water, and moisture are mediated, and for lubrication system, the grease and oil as well as fatty acids type lubricants could be applied on the surfaces to allow slide easily over each other. It is well known that there are possibly four regimes of lubrication effect in an interface with fluid: (i) static friction, (ii) boundary lubrication, (iii) mixed lubrication (partial fluid and partial solid), and (iv) full fluid lubrication. Figure 3.15 gives the schematic of generalized Stribeck curve showing the possible regimes.

Consider the contact sliding interface with a liquid film. To discuss the four possible lubrication regimes, let's define a ratio to quantify the relative thickness of liquid and the roughness. λ is defined as the ratio of the mean liquid film thickness at interface to the combined roughness of both surfaces,

$$\lambda = h_0/R_q \tag{3.86}$$

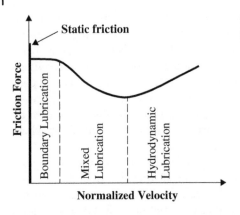

Figure 3.15 Schematic of generalized Stribeck curve.

in which h_0 is the relative thickness of liquid, R_q is the combined average rms roughness of the two surfaces.

Static friction is the tangential force that must be overcome to initiate sliding of one object over another, while kinetic friction is the tangential force that must be overcome to maintain sliding. Strictly speaking, static friction is not a friction force but a threshold of static force. Its presence implies that the objects have locked together into a local energy minimum that must be overcome before motion by the external force.

Different mechanisms have been suggested for static friction such as geometric interlocking, elastic instabilities, interlocking mediated by so-called third bodies, plastic deformation, and crack propagation, just to name a few. All these mechanisms may well be applicable, although each one under different circumstances. Basically, the first regime does not depend on velocity. The static contact allows the asperity junctions to deform elastically and plastically and yields static bonding. A traction force is needed to breakaway the bond, or static friction. There is usually a pre-sliding displacement associated with the breakaway process. The ratio of the breakaway force to normal force is the coefficient of friction. The contacts are compliant in both normal and tangential directions. The junction in the surfaces of the continuum system is like an equivalent spring k. Under the traction, the pre-sliding displacement is linearly proportional to the applied force until to a critical value equal to the static friction then breakaway occurs, the transition to sliding is actually not abrupt. Besides the solid–solid bond, meniscus due to liquid mediation also contributes to the static friction.

The second possible regime is boundary lubrication which occurs at very low sliding velocity for $\lambda < 1$. The velocity is not sufficient to establish a film between the solid surfaces, and the liquid lubrication is not critical which does not dominant friction. The applied load is actually carried by the surface asperities, and the friction depends upon the lubrication properties of the molecules on the surfaces. Coefficients of friction between 0.06 and 0.1 are typical when a low shear strength boundary film is present. If no such film is present, then coefficients ranging between 0.2 and 0.4 can be exhibited, even rising as high as 1.0 in some cases. In this regime, atomically flat surfaces could be separated by a few molecular layers of liquid, however, the behavior of the interface are qualitatively different from the bulk viscosity which is traditionally associated with lubricants film. The interfacial material can pack into a solid-like structure due to its confinement and exhibit properties such as a finite yield stress and stick slip. The boundary lubrication layer is basically solid-like. There is shearing in the liquid in boundary lubrication due to solid-to-solid contact. Boundary lubrication is basically a process of shear in a solid, despite liquid film involved.

The third possible regime is mixed lubrication regime, it is approximately characterized by $1 \leq \lambda < 3$. In these contacts, part of the load is carried by the liquid, and part by the interacting asperities. The rubbing together of asperities can increase friction, and this can also be minimized by the compounds adhering to the surfaces or the tribolayer in the interface. The interface is partially supported by hydrodynamic force and partially by asperity contact force. Some liquid is expelled by pressure, but viscosity or wetting effect prevents all of the liquid from escaping and thus a film is formed. The friction process could be dominated by the interaction of liquid viscosity, motion speed, pressure, and contact geometry. For this case, the greater viscosity or the higher motion velocity, the thicker the fluid film will be, therefore the lower the cof. The surface roughness, asperity size, and orientation have essential impact on the characteristics of formed films. In some other cases, the friction process could be determined by the tribolayer, speed, pressure, and contact geometry. In this case, the surfaces are so close that the liquid viscosity is relatively not significant. It is the liquid physical and chemical interaction with surface dominates the friction. The mixed lubrication usually has liquid film thickness in the range of 30 nm–3 μm. In tribology practice, the additives are usually used to provide the desired properties together with base liquid that needs low surface tension and a low contact angle. The additives are usually adsorbed on surface or react with it to form monolayers with low shear strength, therefore reduce friction as bottom line to protect the surface wear.

The possible regime four is the full fluid lubrication where the solid-to-solid contact is eliminated. The process is governed by either the hydrodynamic or elasto-hydrodynamic lubrication characterized by $3 < \lambda < 10$. In this situation the surfaces are kept apart by a pressurized fluid. The clearance space is much larger than the average surface roughness, and therefore the surfaces can be considered smooth. The pressurization of the fluid is usually achieved by external means in hydrostatic bearings, but is accomplished in hydrodynamic contacts by the relative motion and geometry of the surfaces.

3.3.3.2 Unsteady Liquid-Mediated Friction

In this section we only discuss the first three possible regimes of friction and lubrication in the interface, which are most likely associated with the friction-vibration interaction problems in engineering.

For liquid mediated interface, the static friction or stiction could exhibit strong time-dependent: the static friction increase with rest time remarkably. The empirical model can also be represented by previous formula Eq. (3.84), the power law or divided formula can also be employed. In some cases the static friction starts to rapidly increase after a certain rest time. The long term friction after days dwell could be 10 times higher than the static friction without dwell or having short dwell. Moreover, the higher acceleration, the higher the static friction due to viscous effects for liquid mediated friction.

Generally, there are several possible mechanisms allow static friction force to increase with the contact time: the increasing of the area of real contact with the time of stationary contact; the re-distribution and accumulation of the interface liquid, the micro-flow of lubricant around the toe-dipping regimes, and chain inter-diffusion. Moreover, if the liquid film exists dynamical phase transitions from fluid state during slip to a solid state during stick whose formation is a nucleation process, then the static friction force could also increase with the time of stationary contact.

Figure 3.16 shows the measured static friction coefficient as a function of dwell time for different lubricants.

Many engineering wet interface or liquid mediated interfaces exhibit the unsteady mixed lubrication. The boundary and mixed lubrication is important for friction-vibration interactions because they are usually associated with the stiction and the negative slope of

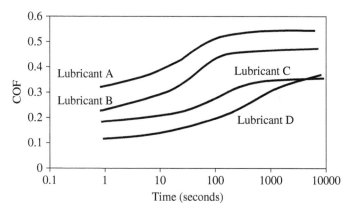

Figure 3.16 Static friction coefficient as a function of dwell time for different lubricants.

friction vs. velocity curve. For most wet interface, the measured static friction is almost always higher than kinetic frictions, so the stick–slip and/or negative damping could concerns. In mixed lubrication regime, friction is a function of velocity because the physical process of shear in the junction changes with velocity and the liquid film has hydrodynamic effect. Details of the velocity-dependent friction feature depend upon the degree of boundary lubrication and the details of mixed lubrication.

Fluid friction arises from the shear of fluid film within contact, can be represented as [100].

$$P_f = P_f(V, P, \eta_i, \alpha, h_0, E^*, r', T, b) \tag{3.87}$$

In which in which $V, P, \eta_i, \alpha, h_0, E^*, r', T, b$ are respectively the velocity, load, fluid viscosity, pressure-viscosity coefficient, film thickness, effective elastic modulus, combined radius of curvature, temperature, film length respectively. On the other hand, the solid friction force can be written, in functional form as,

$$P_s = P_s(P, h_0, E^*, r', \sigma, b) \tag{3.88}$$

in which σ is the combined surface roughness. To model the total friction we add the two forces and divided by the total normal load to acquire friction coefficient.

$$\mu = \frac{P_f + P_s}{P} = \mu(V, P, \eta_i, \alpha, h_0, E^*, r', T, \sigma, \tau_s, b) \tag{3.89}$$

For mixed lubrication regime, liquid viscosity could have insignificant effect on the friction characteristics, the interface load is dominated by asperity contacts. However, despite the liquid film does not control the friction, tribolayer could play a key role to distinguish friction properties between wet friction and dry friction, this could be the case for many wet interface.

To acquire some insight, let's consider contact with hydrodynamic lubrication. The minimum film thickness for smooth surfaces with hydrodynamic lubrication can be calculated from the following formula:

$$h_0 = \frac{c\mu_0 R V}{P} \tag{3.90}$$

where c is constant, P is the normal load per unit width of the contact, R is the relative radius of curvature of the contacting surfaces, E' is the effective elastic modulus, μ_0 is the lubricant viscosity at inlet conditions, and V is the relative surface velocity.

To estimate whether the interface is under lubrication boundary lubrication, we can calculate the specific film thickness or the lambda ratio. For this case, the total normal load is shared between the asperity load and the film load.

Consider the particular lubricated concentrated contact, both the contacting asperities and the lubricating film contribute to supporting the load. Thus

$$P = P_f + P_s \tag{3.91}$$

where P is the total load, P_l is the load supported by the liquid film and P_s is the load supported by the contacting asperities. Load P_s supported by the contacting asperities results in the asperity pressure p_a, given by

$$p_a = \frac{4}{3}(\eta R_p \sigma_p) E^* (\sigma_p / R_p)^{1/2} F_{3/2}(d_e / \sigma) \tag{3.92}$$

where $F_m(D)$ is a statistical function in the Greenwood-Williamson model of contact between two real surfaces, R is the relative radius of curvature of the contacting surfaces, E^* is the effective elastic modulus, N is the asperity density, r is the average radius of curvature at the peak of asperities, σ is the standard deviation of the peaks, and de is the equivalent separation between the mean height of the peaks and the flat smooth surface. For the case $\lambda \approx 1$, the liquid pressure to total pressure is approximated by

$$p_l = \frac{1}{\lambda} \left(\frac{h_0}{h} \right)^\alpha p \tag{3.93}$$

α is a constant, p is the total pressure, h is the mean thickness of the film between two actual rough surfaces, and h_0 is the film thickness with smooth surfaces.

In liquid mediated interface, there exist strong friction delay-hysteresis phenomena. If there is a change in velocity, the corresponding change in friction has a delay. The friction memory can be attributed to the friction level lags a change in system state [86]. The hysteresis is the separation in frictions levels during acceleration and deceleration.

An alternative to the state variable models described in previous section is the pure time lag model. In lubricated contacts, simple time delay better describes the effect.

$$F(t) = F_{vel}(\dot{x}(t - \Delta t)) \tag{3.94}$$

in which $F(t)$ is the instantaneous friction force, $F_{vel}(\cdot)$ is friction as a function of steady state velocity, and Δt is the lag parameter, and it is the time by which a friction lags a change in velocity. The lag increases with the increase of liquid viscosity and contact force.

3.3.3.3 Negative Slope of Friction-Velocity Curve

The liquid mediated friction occurs in many engineering applications. The liquid mediated friction is usually smaller than the dry friction, but there are many reversal results in which the liquid mediated friction is higher than dry friction. Moreover, the trend of velocity-dependent feature of friction is usually determined by the competing of many factors and could have larger variations. Next we illustrate this by discussing several examples in engineering.

Figure 3.17 plots the cof as a function of velocity for an interface with three different lubricant fluids applied. A is considered to be favorable. B and C may be susceptibly vulnerable to self-excited vibrations. The optimization of the fluid is mainly achieved by regulating the additives in the base liquid. On the other hand, the similar profile group could be attained by using same fluid but with different solid friction materials.

3.3.4 Friction Models

It is in general unlikely to study sliding friction dynamics directly using "first-principle" methods such as molecular dynamic calculations. It is also unlikely to study contact sliding dynamics directly using multi

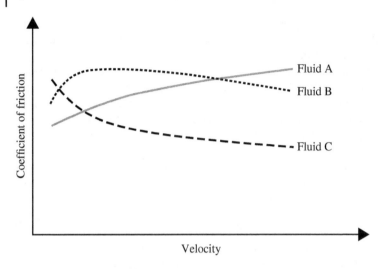

Figure 3.17 Schematic of friction-velocities curves of three fluids.

physics based models where the Newton's force and motion equation, asperity contact mechanics, liquid film dynamics, heat transfer, etc. are all included. The multi physics based models are suitable for fundamental understanding and applications interpretation as well as in-depth investigation and research. However, the model is computationally expensive and too complicated to be implemented compared with empirical models, limiting its applications in engineering and science. For engineering applications, it is usually needed to model friction of construct friction laws which can be used directly in Newton's equation of motion to predict the dynamic performance of the sliding macroscopic bodies [167–177].

It is usually feasible to construct empirical friction laws based on phenomenological observations to predict the sliding behavior of macroscopic bodies. This kind of models is based on macroscopic quantities observed, the displacement, the state variables, load, temperature, and their history.

When we consider elastic continuum vibrations, the friction description could be from as simple as Coulomb friction law to micro-level surface tribological description detailed in Chapter 3. The former could be too simple to capturing dynamic friction phenomena sufficiently; the later could be too complicated to be brought into the differential equation of dynamics motion. There have been many efforts in science and engineering communities to develop trade-off friction law for various problems, and numerous friction laws have been developed with phenomenological expressions.

The normal component of contact forces relating to the deformation of the contact areas can usually use the Hertz theory for approximation. Consider an interface represented by spherical contact, with effective radii of curvature R. Based on Eqs. (3.9)–(3.12), the contact force $F_N(t)$ can be assumed as a function of elastic deformation in terms of normal displacement y_H,

$$F_N(t) = k_H y_H^{3/2}(t) \tag{3.95}$$

in which,

$$k_H = 4E^* \sqrt{R}/3. \tag{3.96}$$

If the contact is remained, this type of contact introduces nonlinear stiffness to system. If the contact gets lost, then a strong nonlinearity of vibro-impact will occur in the system. On the other hand, the coefficient of friction as a mathematical function of a variety of parameters can be developed by using the fundamental formulations described in last chapter.

The most widely used friction law is about the friction-velocity curve characterizing Stribeck features of real systems. The phenomenological expressions for a friction force vs. slip velocity lead to qualitative results that help characterizing underlying physical phenomena. Polynomial and exponential forms of friction laws are used most commonly for capturing this kind attributes. One of the exponential expressions has the following form,

$$\mu(v_r) = \text{sgn}(v_r)(1 - e^{-\beta_v |v_r|})[1 + (f_r - 1)e^{-\alpha_v |v_r|}], \tag{3.97}$$

where v_r is slip speed, α_v and β_v are constants with f_r representing the ratio of the static to kinetic friction coefficients.

Some applications require a more detailed and more accurate representation of friction to better predict system response and the development of instabilities. In these cases, in addition to describing Stribeck curve, it is also needed to quantify friction and coefficient of friction in terms of static friction transition, hysteresis, and pre-sliding displacement (i.e. displacement that occurs just before a complete slip takes place). The pre-sliding displacement is mainly due to lateral contact elasticity. The increase of static friction with time is mainly due to diffusion processes on the interface. The frictional lag is the effect of the friction force lagging in time behind changes in relative velocity and/or normal load.

In cases of oscillatory relative motion involved, particularly those with small amplitudes, the modeling requires further details of the friction force in terms of its dependence on both displacement and velocity. Velocity reversal and the change of friction with displacement are important in many applications, where the friction force remains constant during sliding, but changes with displacement during each reversal of direction. The slope of the friction change with displacement can be treated as the equivalent linear tangential stiffness of the bonding force. Moreover, some modeling needs to take into account of static friction time dependent characteristics.

The analysis of certain dynamic systems needs to represent friction as a function depending on pressure or temperature in addition to velocity [84, 178–184].

A more general description of friction than Eq. (3.96) is given by the following to include Coulomb friction, viscous friction, static friction, and Stribeck effect,

$$
F = \begin{cases} F(v_r) & \text{if } v_r \neq 0 \\ F_e & v_r = 0 \ \text{and} \ |F_e| < F_s \\ F_s \text{sgn}(F_e) & v_r = 0 \ \text{and} \ |F_e| \geq F_s \end{cases} \tag{3.98}
$$

where $F(v_r)$ is an arbitrary function that capturing the friction profile of Stribeck curve. Different parameterizations have been proposed for $F(v_r)$, one of which is $F(v_r) = N\mu(v_r)$, N is normal force and $\mu(v_r)$ is given in Eq. (4.3). A common form for the friction function is given by [178],

$$
F(v_r) = f_c + (f_s - f_c)e^{-|v_r/v_s|^{\delta_s}} + f_v v_r \tag{3.99}
$$

where f_s, f_c, δ_s are constants characterizing static and dynamic friction, v_s is called Stribeck velocity, and f_v represents viscosity.

Another model is the hyperbolic model given by:

$$
F(v_r, x) = \begin{cases} F(x) = \min(|F_{ex}(x)|, F_s)\text{sgn}[F_{ex}(x)], & v_r = 0, \\ F(v_r) = -\frac{F_s \text{sgn}(v_r)}{1+\delta|v_r|}, & v_r \neq 0 \end{cases}
$$

$$\tag{3.100}$$

in which $F_{ex} = kx$ is the externally applied force and it is displacement dependent. The friction force in the stick phase is limited by the maximum static friction force,

$$
|F(x)| \leq F_s. \tag{3.101}
$$

ADCK (Armstrong-H'elouvry, Dupont, Canudas de Wit and Karnopp) model is a more general model with more parameters [178, 179]. It actually consists of two separate models with one for stiction and one for sliding. The friction is modeled as a stiff spring during stiction,

$$F_l(x) = -k_l x \qquad (3.102)$$

x is the pre-sliding displacements, k_l is the equivalent stiffness. When sliding, the friction is modeled as the following form including the Coulomb friction and viscous friction as well as Stribeck friction,

$$F_f(\dot{x}, t) = \left[F_c + F_v |\dot{x}| + F_s(\gamma, t_2) \frac{1}{1 + (\dot{x}(t - \tau_L)/\dot{x}_s)^2} \right] \mathrm{sgn}(\dot{x}) \qquad (3.103)$$

in which

$$F_s(\gamma, t_2) = F_a + (f_\infty - F_a) \frac{t_2}{t_2 + \gamma} \qquad (3.104)$$

In which F_f is friction force, F_C is Coulomb friction, F_v is viscous force, F_s describes the varying friction level at breakaway or the static friction. The level of the static friction force Fs varies with the dwell time t_2. Force F_a is the magnitude of the Stribeck friction at the end of the previous sliding period; force F_∞ is the magnitude of the Stribeck friction after dwell. γ is an empirical parameter characterizing time dependence of static friction. τ_L is time delay accounting for the desired frictional memory. $\dot{x}_s$ is the characteristic velocity of Stribeck friction.

It is noted that the inherent characteristics of the velocity dependence of friction actually make it difficult to observe this phenomenon accurately in experiment. The friction is typically investigated via steady state sliding experiments in friction tester whose response is an elastic rather than viscous to sudden displacement input. A coefficient of friction of negative slope could eliminate the possibility of steady state sliding as the dynamics of the tester consists of a negative damping and results in unstable oscillations. In other words, the apparent phenomena of a negative slope of friction vs. speed in a steady state sliding experiment could be the result of the interactions between the system dynamics and the complicated constitutive relationship for friction. As we discussed in last chapter, the oscillation of system could reduce interface friction, and system oscillation due to asperity effect is usually proportional to sliding speed, thus the friction speed dependence could be the result of interaction. Because of this type of considerations, rate, and state friction models have been developed to simulate friction dynamics [87, 185]. In these models, the dependence of friction on the relative velocity between the two bodies in contact is modeled using a differential equation. These models include the pre-sliding displacement. The state variable friction models have been developed to quantify dynamic friction behavior including Stribeck friction, rising static friction, and frictional memory.

A widely used nonlinear model characterizing hysteretic behavior in mechanical system is the Bouc-Wen model [186–191]. The model was developed to predict various hardening or softening, and smoothly varying or nearly bilinear hysteretic behavior. This model was extended to characterize the hysteretic behavior as shown in Figure 3.18. The model is restricted to rate-independent hysteretic systems in which velocity-dependent damping effects are nearly negligible. The model has also extended to combine with the Coulomb friction and velocity-dependent damping in order to deal with the rate-dependent damping behavior of some systems like wire cable isolator. In Bouc-Wen model, the interfacial force is assumed to consist of a linear part and a nonlinear part,

$$F(t) = kx(t) + k_n z(t) \tag{3.105}$$

in which the friction force is denoted by $F(t)$, the excitation displacement is denoted by $x(t)$, and k is linear spring stiffness. The nonlinear force is characterized by $z(t)$. The constant k_n is used to scale the

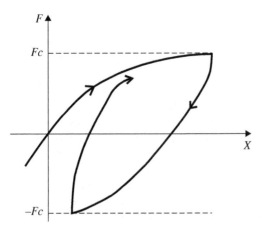

Figure 3.18 Friction force as a function of displacement.

nonlinear force, so that here $z(t)$ has the dimension of displacement. The Bouc-Wen model is represented by the following form,

$$\dot{z}(t) = \alpha \dot{x}(t) - \beta |\dot{x}(t)||z(t)|^{n-1}z(t) - \gamma \dot{x}(t)|z(t)|^n \tag{3.106}$$

where the dot denotes the time derivative, and α, β, γ, n are the model parameters to be determined. The parameter α, β, γ and the exponent n can be estimated by parameter identification techniques based on experimental data.

In solid mechanics, various models have been developed to characterize hysteresis. One of such models is Dahl model [192], which uses the stress–strain curve in classic solid mechanics as starting point for the friction. The stress–strain curve is modeled using the following differential equation:

$$\frac{dF}{dx} = \sigma \left(1 - \frac{F}{F_c} sgn(v_r)\right)^i \tag{3.107}$$

where x is the displacement, F is the friction force, F_c is the Coulomb friction force, σ is the contact stiffness coefficient, v_r is slip speed, and the exponent i determines the shape of the stress–strain curve. Then the following friction model is introduced by including presliding displacement and incorporating tangential compliance,

$$F = \sigma z, \quad \sigma > 0; \quad \dot{z} = \dot{x}\left[1 - \frac{\sigma}{F_c} sgn(\dot{x})z\right]^i \tag{3.108}$$

where z specifies the state of strain in the frictional contact. Friction force is determined for all trajectories. The integer exponent i was used to characterize the transition rate of z so as to attain an optimal experimental fit. For simplification the exponent i is usually assumed to be 1. The steady state version of the above model gives Coulomb friction. Dahl's model is a comparatively simple model capturing many phenomena like hysteresis, but it does not cover Stribeck effect because the model assumes friction depending only on displacement.

LuGre model extended the Dahl model to include the Stribeck effect [85, 193–195]. The LuGre model can captures many aspects of friction such as stick–slip motion. This model with continuous states is interpreted as example of Prandlt's elasto-plastic material model. LuGre model has been used in many complicated engineering problems like tire friction modeling.

The LuGre model is based on an internal variable z which can be understood as the average deflection or tangential strain of the microscopic contact elements, i.e. the asperities or contact zones.

Moreover, some elasto-plastic models been proposed to overcome the drawbacks of LuGre model which exhibits drift for arbitrarily small external force [85]. Schematically, it uses a so-called "bristle" assumption to interpret friction [195]. Assume two proximity contact surfaces with contacting asperities extending from each. The asperities can be represented by small bristles. The friction between the two surfaces is assumed to be caused by many engaged bristles. The portion of friction contributed by each bristle is proportional to the strain of the bristle. When the strain exceeds a certain level, then the bristle engagement or bond is broken. As such, the bristle behaviors like stiff spring with damper, each giving rise to microscopic displacements and restore forces. If the displacement becomes too large, the engaged junctions (stick) break. When this break-away occurs, macroscopic sliding (slip) starts. The friction is thus modeled as the average deflection of the bristles. When a tangential force is applied, the bristles deflect like springs-dashpot. Denoting by z the average bristle deflection as the internal state, the model is given by,

$$F = \sigma_0 z + \sigma_1 \dot{z} + \sigma_2 v_r, \quad \sigma_i > 0, \quad i = 0, 1, 2 \tag{3.109}$$

in which $\sigma_0, \sigma_1, \sigma_2$ are Coulomb, damping, and viscous friction parameters for the tangential compliance, and the $\dot{z}$ is given by,

$$\dot{z} = v_r \left(1 - \frac{\sigma_0}{f_{ss}(v_r)} sgn(v_r) z \right)^i \tag{3.110}$$

σ_0 can be considered as the stiffness of the bristles, and σ_1 is the damping. One more extension of the models allowing a purely elastic regime is given by the following formulations [85, 184],

$$F(z, \dot{z}, v_r, w) = \sigma_0 z + \sigma_1 \dot{z} + \sigma_2 v_r + \sigma_3 w, \quad \sigma_i > 0, \quad i = 0, 1, 2, 3, \tag{3.111}$$

in which

$$\dot{z}(z, v_r) = v_r \left[1 - \alpha(z, v_r) \frac{\sigma_0 sgn(v_r) z}{y_{ss}(v_r)} \right]^i \tag{3.112}$$

where $\alpha(z, v_r)$ is an adhesion map which controls the rate of change of z in order to avoid drift.

The model defines the averaged bristle behavior as a first-order system, with z and $\dot{z}$ being interpreted as the mean bristle displacement and velocity respectively. v_r is the relative velocity between the two bodies in contact. The model suggests that the friction force F results from four components: an elastic term, an internal dissipation term, a viscosity term, and a noise term. The fourth component term $w(t)$ is assumed to be as a pseudo random function of time series representing the random effects or uncertainty of interface like surface roughness. The auxiliary functions z_{ss} and α are defined as [30],

$$y_{ss}(v_r) = [f_c + (f_s - f_c)e^{-(v_r/v_s)^2}] \tag{3.113}$$

$$\alpha(v_r, z) = \begin{cases} \begin{rcases} 0 & |z| < z_{ba} \\ \alpha_m(v_r, z) & z_{ba} < |z| < z_{max}(v_r), \\ 1 & |z| > z_{max}(v_r), \end{rcases} & if\ sgn(v_r) = sgn(z) \\ 0 & if\ sgn(v_r) \neq sgn(z), \end{cases} \tag{3.114}$$

in which the function $\alpha_m(v_r, z)$ is parameterized as

$$\alpha_m(v_r, z) = \frac{1}{2}\left[1 + sin\left(\pi \frac{z - (z_{max}(v_r) + z_{ba})/2}{z_{max}(v_r) - z_{ba}}\right)\right] \tag{3.115}$$

which describes the transition between elastic and plastic behavior. The parameter z_{ba} defines the point where α starts to take nonzero values and is called as breakaway displacement.

In general, for the constant normal stress cases, the generalized friction model has the following form [179–182],

$$F_f(t) = f(V, x_1, x_2, \cdots, x_n) \tag{3.116}$$

$$\dot{x}_i = g_i(V, x_1, x_2, \cdots, x_n), i = 1, 2, \cdots n$$

In which x_i is state variable. This model suggests that a sudden velocity change is unable to create a sudden change in the state, but change its time derivative. The state variables can be found many physical interpretations in applications.

The above models only consider the interfacial tangential direction force and deformation. Oden and Marins introduced a constitutive model for frictional interfaces that represents both the nonlinear normal compliance and sliding resistance of the interface [95, 96]. The constitutive equations consist of two parts: a normal interface law and a friction law. For the normal response, the normal compliance of the interface is modeled as,

$$\sigma_N = c_N a^{m_N} + b_N a^{l_N} \dot{a}, \text{ for } a \geq 0, \tag{3.117}$$

$$\sigma_N = 0, \text{ for } a < 0 \tag{3.118}$$

where σ_N is the normal stress, a is the normal approach on the interface, and $\dot{a}$ is its time derivative. The coefficients c_N, b_N, m_N, l_N depend on the properties of the contacting surfaces and the materials of the two components. The second constitutive equation gives friction law:

$$\text{When } a < 0, \text{ then } \sigma_T = 0 \tag{3.119}$$

When $a \geq 0$, then $|\sigma_T| \leq c_T a^{m_T}$

$$\text{And } \begin{cases} |\sigma_T| < c_T a^{m_T}, & \text{then } \dot{d} = 0 \\ |\sigma_T| = c_T a^{m_T}, & \text{then } \dot{d} = -\lambda \sigma_T \ (\lambda \geq 0) \end{cases} \tag{3.120}$$

where $\dot{d}$ is the sliding velocity calculated as the time derivative of sliding distance, and the index T indicates a direction tangential to the contact surface. The friction force is a function of the normal approach of the two surfaces, which in turn depends on the normal force.

For a given surface profile and response curves, the parameters in Oden-Martins law can be determined and the static coefficient of friction can be estimated. The friction law was also modified to model velocity-dependence, in which the coefficient of friction is given as a function of the relative tangential velocity at the interface,

$$\mu = \left| \frac{\sigma_T}{\sigma_N} \right| = \mu(v_T) \tag{3.121}$$

A further extension of the law is that the coefficient of friction depends on the normal load. The load dependence is often weak for hard materials, but may be appreciable for relatively soft material. For instance, the Eq. (3.72) is a simplified friction model to characterize the relationship between coefficient of friction and interface contact pressure.

In engineering applications, data fitting based empirical models have been used to quantify friction as the function of operation parameters like velocity, pressure, temperature, etc. If the involved database is very big, and the engagement data are directly gathered or indirectly derived from a variety of sources, the data can be simply treated as a large look-up table of friction under various conditions. They can be reduced to a compact mathematical representation through a system identification technique or using a typical static model. Static model is most commonly used, in which friction is measured as a function of applied pressure, slip speed, temperature, and time. More process parameter, denoted as L, as additional variable may be added. Mathematically, the friction can be written as,

$$F = function(p, V, T, L, t) \tag{3.122}$$

In practice, an empirical model is often coupled with an algebraic model.

Based upon certain system identification techniques, assume a certain functional form $f_1, \cdots f_k$ with a set of unknown parameters, $a_{1n}, \cdots a_{in}, i = 1, \cdots k$. Then we can express friction as,

$$F = \sum_k f_i(p, V, T, L, t | a_{i1}, \cdots a_{in}) \tag{3.123}$$

It can be fitted to the measured data, and then it is solved for a friction coefficient set. Many techniques are available to prescribe this functional form and to determine its empirical parameters.

Generally, the modeled friction should be verified and validated. The friction is associated with system dynamics, interface geometry, mechanical, chemical, and physical properties of system.

Consider a dynamic mechanical/structural system consisting of sliding interface and assume the system has dynamic load $P(t)$ as shown in Figure 3.19. The apparent velocity is V_0, $\dot{x}(t)$ is vibration velocity for an arbitrary instant. The normal force is $N(t)$ and the tangential resistance force $F(t)$ is the friction force. The normal force comprised of both the dynamic load due to the system and the normal component of any adhesive forces in interface.

For a general dynamic system, obviously $F(t)$ is also an internal response or responsive force which depends on entire system dynamics (for example, through $N(t)$ and $\dot{x}(t)$) as well as the interface state and parameters (adhesion, wear, fracture, interlock, interface liquid films, etc.). The interface affect system dynamics through $N(t)$ and $F(t)$.

The interface forces are not fully determined in prior by interface parameters. In principle, the interface forces are time-varying response of the dynamic system.

In most engineering applications, the concept of coefficient of friction $\mu = \overline{F}(t)/\overline{N}(t)$ has been used to simplify the analysis. In complex dynamics systems, the interface forces as dynamic property or time-varying response of the dynamic system need to be quantified,

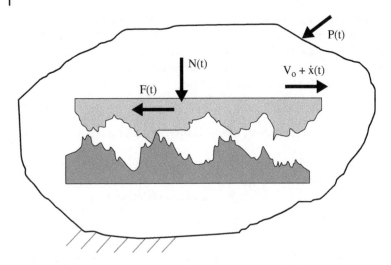

Figure 3.19 Schematic of a dynamical system consisting of sliding interface.

verified, and validated based on measured dynamics response. This is usually conducted by a model-based identification from measurements [196–205].

A comparatively new technique for system identification of friction and contact model is based on an artificial neural network [206–209]. The artificial neural network is a tool for general pattern recognition. It empirically identifies input–output relationships of an unknown system based upon available data. The neural net is especially applicable for a non-linear system-identification or used as system-modeling tool when the physical nature of a target system is not well-understood. Next we briefly illustrate basic neural net structure and operations for the friction component modeling. The artificial neural net can be considered as a "black box" that accurately recreates system friction behaviors even if the system is indescribable with first principles. A two-layer feedforward neural net has the prescribed functional form as follows [206],

$$y_k = f_2 \left\{ \sum_j W_{jk} \left[f_1 \left(\sum_i W_{ij} x_i \right) \right] \right\} \tag{3.124}$$

where i = index for input variables, j = index for 1st layer neurons, and k = index for output variables. Neural net weights (W_{ij}, W_{ik}) represent unknown model parameters. While a set of data flows through the net from left to right, a connector multiplies data with a weight. Then, a node carries out summation ($\sum$) and a simple linear or non-linear operation (f_1, f_2) such as Sigmoid transformation. It is mathematically proven that this simple two-layer neural net is capable of capturing input to output relationships of any continuous functions when connector weights are suitably selected. Thus, a key to construct a successful neural net friction model is to optimize a set of weights using the data from a target system. This process is often called neural net training and is analogous to determining coefficients in a conventional regression or system identification technique.

References

1 Johnson, K.L. (1985). *Contact Mechanics*. Cambridge: Cambridge University Press.

2 Bowden, F.P. and Tabor, D. (1964). *Friction and Lubrication of Solids*. Oxford: Clarendon Press.

3 Rabinowicz, E. (1995). *Friction and Wear of Materials*, 2e. New York: Wiley.

4 Bhushan, B. (1999). *Principles and Applications of Tribology*. New York: Wiley.

5 Stolarski, T.A. (1990). *Tribology in Machine Design*. Woburn, MA: Butterworth Heinemann.

6 Blau, P.J. (1986). *Friction Science and Technology*. New York: Marcel Dekker.

7 Israelachvili, J.N. (1991). *Intermolecular and Surface Forces*. London: Academic Press.

8 Persson, B.N.J. (1972, 1998). *Sliding Friction: Physical Principles and Applications*. Heidelberg: Springer.

9 Moore, D.F. (1999). *The Friction and Lubrication of Elastomers*. Oxford: Pergamon Press.

10 Bhushan, B. *Handbook of Micro/Nanotribology*, 2e. Boca Raton, FL: CRC.

11 Bhushan, B. and Gupta, B.K. (1991). *Handbook of Tribology: Materials, Coatings and Surface Treatments*. New York: McGraw Hill.

12 Wiesendanger, R. *Scanning Probe Microscopy and Spectroscopy: Methods and Applications*. Cambridge, UK: Cambridge University Press.

13 Adamson, A.W. (1994). *Physical Chemistry of Surfaces*, 5e. New York: Wiley.

14 Sheng, G., and Wang Q. 1990. Brake NVH Technology, AMC report, 2002–82.

15 Fischer-Cripps, A.C. (1999). *Introduction to Contact Mechanics*. Springer.

16 Blau, P.J. (2001). The significance and use of the friction coefficient. *Tribol. Int.* 34: 585.

17 Bhushan, B., Israelachvili, J.N., and Landman, U. (1995). Nano tribology, wear and lubrication at the atomic scale. *Nature, London* 374: 607.

18 He, G., Muser, M.M., and Robbins, M.O. (1999). Adsorbs layers and the origin of static friction. *Science* 284: 1650.

19 Gerde, E. and Marder, M. (2001). Friction and fracture. *Nature, London* 413: 285.

20 Ludema, K.C. (1996). *Friction, Wear, and Lubrication: A Textbook in Tribology*. Boca Raton: CRC Press.

21 Hutchings, I.M. (1992). *Tribology: Friction and Wear of Engineering Materials*. Boca Raton: CRC Press.

22 Singer, I.L. and Pollack, H.M. (1992). *Fundamentals of Friction: Macroscopic and Microscopic Processes*. Dordrecht: Kluwer.

23 Sokolo, J.B. (1990). Theory of energy dissipation in sliding crystal surfaces. *Phys. Rev. B* 42 (760): 6745.

24 Bruch, L.W., Cole, M.W., and Zaremba, E. (1996). *Physical Adsorption: Forces and Phenomena*. Oxford: Clarendon Press.

25 Mate, C.M., McClelland, G.M., Erlandsson, R. et al. (1987). Atomic-scale friction of a tungsten tipon a graphite surface. *Phys. Rev. Lett.* 59: 1942.

26 Greenwood, J.A. and Williamson, J.B.P. (1966). Contact of nominally flat surfaces. *Proc. R. Soc. London, Ser. A* 295: 300.

27 Johnson, K.L., Kendall, K., and Roberts, A.D. (1971). Surface energy and the contact of elastic solids. *Proc. R. Soc. London, Ser. A* 324: 301.

28 Tolstoi, D.M. (1967). Significance of the normal degree of freedom and natural normal vibrations in contact friction. *Wear* 10 (3): 199.

29 Bhushan, B. (1996). Contact mechanics of rough surfaces in tribology: single asperity contact. *Appl. Mech. Rev.* 49: 275.

30 Bhushan, B. (1998). Contact mechanics of rough surfaces in tribology: multiple asperity contact. *Tribol. Lett.* 4: 1.

31 Bhushan, B. and Majumdar, A. (1992). Elastic-plastic contact model of bifractal surfaces. *Wear* 153: 53.

32 Peng, W. and Bhushan, B. (2001). A numerical three-dimensional model for the contact of layered elastic/plastic solids with rough surfaces by variational principle. *ASME J. Tribol.* 123: 330.

33 Yu, M. and Bhushan, B. (1996). Contact analysis of three-dimensional rough surfaces under frictionless and frictional contact. *Wear* 200: 265.

34 Chilamakuri, S.K. and Bhushan, B. (1998). Contact analysis of non-Gaussian random surfaces. *Proc. Inst. Mech. Eng. Part J: J. Eng. Tribol.* 212: 19.

35 Kogut, L. and Etsion, I. (2004). A static friction model for elastic-plastic contacting rough surfaces. *Trans. ASME J. Tribol.* 126: 34.

36 Etsion, I. and Front, I. (1994). A model for static sealing performance of end face seals. *Tribol. Trans.* 37: 111.

37 Chang, W.R., Etsion, I., and Bogy, D.B. (1988). Static friction coefficient model for metallic rough surfaces. *ASME J. Tribol.* 110: 57.

38 Chang, W.R., Estion, I., and Bogy, D.B. (1987). An elastic-plastic model for the contact of rough surfaces. *ASME J. Tribol.* 101: 15.

39 Kogut, L. and Etsion, I. (2003). A semi-analytical solution for the sliding inception of a spherical contact. *ASME J. Tribol.* 125: 499.

40 Chang, W.R., Etsion, I., and Bogy, D.B. (1988). Adhesion model for metallic rough surfaces. *ASME J. Tribol.* 110: 50.

41 Kogut, L. and Etsion, I. (2003). Adhesion in elastic-plastic spherical micro-contact. *J. Colloid Interface Sci.* 261: 372.

42 King, R.B. and O'Sullivan, T.C. (1987). Sliding contact stresses in a two-dimensional layered elastic half-space. *Int. J. Solids Struct.* 23: 581.

43 Komvopoulos, K. (1989). Elastic-plastic finite element analysis of indented layered media. *ASME J. Tribol.* 111: 430.

44 Yan, W. and Komvopoulos, K. (1998). Contact analysis of elastic–plastic fractal surfaces. *J. Appl. Phys.* 84: 3617.

45 Komvopoulos, K. and Choi, D.H. (1992). Elastic finite element analysis of multiasperity contact. *ASME J. Tribol.* 114: 823.

46 Kral, E.R. and Komvopoulos, K. (1997). Three-dimensional finite element analysis of subsurface stress and strain fields due to sliding contact on an elastic-plastic layered medium. *ASME J. Tribol.* 119: 332.

47 Malvern, L.E. (1969). *Introduction to the Mechanics of a Continuous Medium*. Prentice-Hall.

48 Mao, K., Bell, T., and Sun, Y. (1997). Effect of sliding friction on contact stresses for multilayered elastic bodies with rough surfaces. *ASME J. Tribol.* 119: 476.

49 Merriman, T. and Kannel, J. (1989). Analyses of the role of surface roughness on contact stresses between elastic cylinders with and without soft surface coating. *ASME J. Tribol.* 111: 87.

50 Nogi, T. and Kato, T. (1997). Influence of a hard surface layer on the limit of elastic contact-part I: analysis using a real surface model. *ASME J. Tribol.* 119: 493.

51 Peng, W. and Bhushan, B. (2000). Modeling of surfaces with bimodal roughness distribution. *Proc. Inst. Mech. Eng. Part J: J. Eng. Tribol.* 214: 459.

52 Polonsky, A. and Keer, L.M. (1999). A numerical method for solving rough contact problems based on the multi-level multi-summation and conjugate gradient techniques. *Wear* 231: 206.

53 Polonsky, A. and Keer, L.M. (2000). Fast methods for solving rough contact problems: a comparative study. *ASME J. Tribol.* 122: 36.

54 Bush, A.W., Gibson, R.D., and Keogh, G.P. (1976). The limit of elastic deformation in the contact of rough surfaces. *Mech. Res. Commun.* 3: 169.

55 Pollock, H.M. (1992). Surfaces forces and adhesion. In: *Fundamentals of Friction: Macroscopic and Microscopic Processes* (ed. I.L. Singer and H.M. Pollock), 77. Dordrecht: Kluwer Academic Publishers.

56 Ogilvy, J.A. (1992). Numerical simulation of elastic-plastic contact between anisotropic rough surfaces. *J. Phys. D* 25: 1798.

57 Rymuza, Z. (1996). Energy concept of the coefficient of friction. *Wear* 199: 187.

58 Nayak, P.R. (1971). Random process model of rough surfaces. *J. Lubr. Technol.* 93: 398.

59 Yu, N. and Polycarpou, A.A. (2002). Contact of rough surfaces with asymmetric distribution of asperity heights. *ASME J. Tribol.* 124: 367.

60 Sheng, G., Liu, B., and Hua, W. (2001). Discussion "the stresses induced by dynamic load head-disk contacts and the parameter extrapolation for design", ASME transaction. *J. Tribol.* 123: 655.

61 Sheng, G., Chen, Q.S., Hua, W. et al. An experimental study of dimple separations and head-disk impacts of negative pressure slider in unload process. *IEEE Trans. Magn.* 37: 1859.

62 Sheng, G. and Zhang, J. (2001). An experimental and theoretical investigation of disk damage caused by head-disk impact in loading process. *IEEE Trans. Magn.* 37: 1863.

63 Hua, W., Liu, B., Li, J. et al. (2001). Further studies of unload process with a 9D model. *IEEE Trans. Magn.* 37: 1855.

64 Sheng, G., Hua, W., and Zhang, J. (2001). Head-disk impact stress in dynamic loading process and the extrapolation of parameters for slider rounding and interface durability. *J. Inf. Storage Process. Syst.* 3: 4.

65 Sheng, G., Liu, B., and Hua, W. (2000). An nonlinear dynamics theory for modeling slider-air bearings in hard disk drives. *J. Appl. Phys.* 87 (9): 6173.

66 Sheng, G., Hua, W., Chen, Q.S. et al. (2000). Experimental and analytical study of head-disk impact and dynamics of negative pressure slider in unload process. *J. Inf. Storage Process. Syst.* 2: 281.

67 Hua, W., Liu, B., Sheng, G. et al. (2000). Investigations of disk surface roughness on the dynamic performance of proximity recording slider. *J. Magn. Magn. Mater.* 209: 163.

68 Sheng, G., Liu, B., Hua, W. et al. (2000). Stable interface concept and design for nano-meter spacing magnetic recording. *J. Magn. Magn. Mater.* 209: 160.

69 Sheng, G., Liu, B., Hua, W. et al. (1999). A theoretical model of acoustic emission sensing process and the experimental investigations for near-contact and contact interfaces in magnetic recording system. *J. Appl. Phys.* 85 (8): 5621.

70 Sheng, G. and Liu, B. (1999). A theoretical model of slider-disk interaction and acoustic emission sensing process for studying interface phenomena and estimating unknown parameters. *Tribol. Lett.* 6: 233.

71 Sheng, G., Liu, B., and Hua, W. (1999). Structure and mechanics study of slider design for 5–15 nm head-disk spacing. *IEICE Trans. Electron.* E82-C (12): 2125.

72 Sheng, G., Liu, B., and Hua, W. (1999). A micro-machined dual slider-suspension for near-contact and contact recording. *IEEE Trans. Magn.* 35 (5): 2472.

73 Liu, B., Sheng, G., Hua, W. et al. (1999). A dual-stage slider-suspension design for contact recording. *J. Appl. Phys.* 85 (8): 5609.

74 Hua, W., Liu, B., and Sheng, G. (1999). Probability model and its application on the interaction of nano meter spaced slider-disk interface. *IEICE Trans. Electron.* E82-C (12): 2139.

75 Zhu, Y.L., Liu, B., and Sheng, G. (1999). Sensitivity study of 40 Gb/In^2 recording system to the fluctuation of head-disk spacing. *IEEE Trans. Magn.* 35 (5): 2241.

76 Hua, W., Liu, B., Sheng, G. et al. (1999). Probability model for slider-disk interaction. *J. Inf. Storage Process. Syst.* 1: 273.

77 Hua, W., Liu, B., and Sheng, G. (1999). Disk roughness and its influence on the performance of proximity recording sliders. *IEEE Trans. Magn.* 35 (5): 2460.

78 Liu, B., Sheng, G., Chen, Q. et al. (1998). Study of clock head-disk interface failure mechanism in servo-writing process. *IEEE Trans. Magn.* 34 (4): 1723.

79 McCool, J.I. (1987). The distribution of microcontact area, load, pressure and flash temperature under the greenwood–Williamson model. *ASME J. Tribol.* 87, Trib-25.

80 McCool, J.I. (2000). Extending the capability of the greenwood–Williamson microcontact model. *ASME J. Tribol.* 122: 496.

81 Deng, K. and Ko, W.H. (1992). A study of static friction between silicon and silicon compounds. *J. Micromech. Microeng.* 2: 14.

82 Smith, D.P. (1999). Tribology of the belt-driven data tape cartridge. *Tribol. Int.* 31 (8): 465–477.

83 Alciatore, D.G. and Traver, A.E. (1995). Multipulley belt drive mechanics: creep theory vs shear theory. *Trans. ASME, J. Mech. Des.* 117: 506.

84 Polycarpou, A.A. and Soom, A. (1995). Application of a two-dimensional model of continuous sliding friction to stick-slip. *Wear* 181/183: 32.

85 Dupont, P., Hayward, V., Armstrong, B. et al. (2002). Single state elastoplastic friction models. *IEEE Trans. Autom. Control* 47 (5).

86 Armstrong-Helouvry, B., Dupont, P., and Canudas de Wit, C. (1994). A survey of models, analysis tools and compensation methods for the control of machines with friction. *Automotica* 30 (7): 1083.

87 Dieterich, J. (1979). Modelling of rock friction: 1. Experimental results and constitutive equations. *J. Geophys. Res.* 84: 2161.

88 Ruina, A.L. (1983). Slip instability and state variable friction laws. *J. Geophys. Res.* 88: 10359.

89 Hess, D.P. 1996. *Interaction of Vibration and Friction at Dry Sliding Contacts, Dynamics with Friction*, ed. By Guran, A., Pfeiffer F., and Popp, K., World Scientific Publishing, New Jersey, 1, part II.

90 Urbakh, M., Klafter, J., Gourdon, D. et al. (2004. The nonlinear nature of friction). *Nature* 430: 29.

91 Zhao, Y., Maietta, D.M., and Chang, L. (2000). An asperity microcontact model incorporating the transition from elastic deformation to fully plastic flow. *ASME J. Tribol.* 122: 86.

92 Greenwood, J.A. and Wu, J.J. (2001). Surface roughness and contact: an apology. *Meccanica* 36 (6): 617.

93 Chang, W.R. (1997). An elastic-plastic contact model for a rough surface with and ion-plated soft metallic coating. *Wear* 212: 229.

94 Berthe, D. and Godet, M. (1973). A more general form of Reynolds equation – application to rough surfaces. *Wear* 27: 345.

95 Oden, J.T. and Martin, J.A.C. (1985). Models and computational methods for dynamic friction phenomena. *Comput. Methods Appl. Mech. Eng.* 52: 527.

96 Martin, J.A.C., Oden, J.T., and Simoes, F.M.F. (1990). A study of static and kinetic friction. *Int. J. Eng. Sci.* 28: 29.

97 Tworzydlo, W.W., Cecot, W., Oden, J.T. et al. (1998). Computational micro- and macroscopic models of contact and friction: formulation, approach and applications. *Wear* 220: 113.

98 Polycarpou, A.A. and Etsion, I. (1998). Static friction of contacting real surfaces in the presence of sub-boundary lubrication. *ASME J. Tribol.* 120: 296.

99 Polycarpou, A.A. and Etsion, I. (1998). Comparison of the static friction sub-boundary lubrication model with experimental measurements on thin-film disks. *Tribol. Trans.* 41: 217.

100 Polycarpou, A. and Soom, A. 1996 Modeling unsteady lubricated friction, 197 in *Dynamics with Friction*, ed. By Guran, A., Pfeiffer F., and Popp, K., New Jersey, World Scientific Publishing.

101 Chang, L. (1995). A deterministic model for line–contact partial elasto–hydrodynamic lubrication. *Tribol. Int.* 28: 75.

102 Wu, C.W. and Zheng, L.Q. (1989). An average Reynolds equation for partial film lubrication with a contact factor. *J. Tribol.* 111: 188.

103 Christensen, H. and Tønder, K. (1973). The hydrodynamic lubrication of rough journal bearings. *J. Lubr. Technol.* 95: 166.

104 Timoshenko S., and Woinowsky-Krieger, S. *Theory of Plates and Shells.* McGraw-Hill Book Company, 2nd edition, 1987.

105 Wang, Q. and Cheng, H.S. (1995). A mixed lubrication model for journal bearings with a thin soft coating – part I: contact and lubrication analysis. *Tribol. Trans.* 38 (3): 654.

106 Wu, J.J. (2001). The properties of asperities of real surfaces. *ASME J. Tribol.* 123: 872.

107 Zhai, X. and Chang, L. (1998). An engineering approach to deterministic modeling of mixed-film contacts. *Tribol. Trans.* 41 (3): 327.

108 Francis, H.A. (1977). Application of spherical indention mechanics to reversible and irreversible contact between rough surfaces. *Wear* 45 (261).

109 Persson, B.N.J. (1998). On the theory of rubber friction. *Surf. Sci.* 401: 445.

110 Person, B.N.J. (1999). Sliding friction. *Surf. Sci. Rep.* 33: 83.

111 Nielsen, L.E. and Landel, R.F. (1991). *Mechanical Properties of Polymers and Composites*. New York: Marcel Dekker.

112 Myshkin, N.K., Petrokovets, M.I., and Kovalev, A.V. (2005). Tribology of polymers: adhesion, friction, wear, and mass-transfer. *Tribol. Int.* 38: 910.

113 Friedrich, K., Lu, Z., and Hager, A.M. (1995). Recent advances in polymer composites tribology. *Wear* 190: 139.

114 Zhao, Z. and Bhushan, B. (1996). Effect of bonded-lubricant films on the tribological performance of magnetic thin-film rigid disks. *Wear* 202: 50.

115 Tian, H. and Matsudaira, T. (1993). The role of relative humidity, surface roughness and liquid build-up on static friction behavior of the head/disk interface. *ASME J. Tribol.* 115: 28.

116 Tian, X. and Bhushan, B. (1996). The micro-meniscus effect of a thin liquid film on the static friction of rough surfaces contact. *J. Phys. D: Appl. Phys.* 29: 163.

117 Bhushan, B., Kotwal, C.A., and Chilamakuri, S.K. (1998). Kinetic meniscus model for prediction of rest stiction. *ASME J. Tribol.* 120: 42.

118 Gao, C., Tian, X., and Bhushan, B. (1995). A meniscus model for optimization of texturing and liquid lubrication of magnetic thin film rigid disks. *Tribol. Trans.* 38: 201.

119 Yufeng, L. and Talke, F.E. (1990). *A model for the effect of humidity on stiction of the head/disk interface, Tribology and Mechanics of Magnetic Storage Systems* (ed. B. Bhushan) SP–29, STLE, Park Ridge, III, 79.

120 Tasy, N., Sonnenberg, T., Jansen, H. et al. (1996). Stiction in surface micromachining. *J. Micromech. Microeng.* 6: 385.

121 Mastrangelo, C.H. and Hsu, C.H. (1993). Mechanical stability and adhesion of microstructures under capillary forces—part I: basic theory. *J. Microelectromech. Syst.* 2: 33.

122 Mastrangelo, C.H. and Hsu, C.H. (1993). Mechanical stability and adhesion of microstructures under capillary forces—part II: experiments. *J. Microelectromech. Syst.* 2: 44.

123 Israelachvili, J.N. and Tabor, D. (1972). The measurement of van der Waals dispersion forces in the range 1.5 to 130 nm. *Proc. R. Soc. London, Ser. A* 331: 19.

124 Matthewson, M.J. and Mamin, H.J. (1988). Liquid mediated adhesion of ultra-flat solid surfaces. *Mater. Res. Soc. Symp. Proc.* 119 (87).

125 Li, Y., Trauner, D., and Talke, F.E. (1990). Effect of humidity on stiction and friction of the head/disk interface. *IEEE Trans. Magn.* 26: 2487.

126 Raman, V., Tang, W.T., Jen, D. et al. (1991). The dependence of stiction and friction on roughness in thin film magnetic recording disks. *J. Appl. Phys.* 70: 1826.

127 Decker, L., Frank, B., Suo, Y. et al. (1999). Physics of contact angle measurement. *Colloids Surf., A* 156: 177.

128 Sedev, R.V., Budziak, C.J., Petrov, J.G. et al. (1993). Dynamic contact angles at low velocities. *J. Colloid Interface Sci.* 159: 392.

129 Kwok, D.Y., Budziak, C.J., and Neumann, A.W. (1995). Measurements of static and low-rate dynamic contact angles by means of an automated capillary rise technique. *J. Colloid Interface Sci.* 173: 143.

130 Van Oss, J. (1994). *Interfacial Forces in Aqueous Media*. New York: Marcel Dekker.

131 Egorov, A.G., Kornev, K.G., and Neimark, A.V. (2003). Meniscus motion in a prewetted capillary. *Phys. Fluids* 15 (10).

132 Kornev, G., Shingareva, I.K., and Neimark, A.V. (2002). Capillary condensation as a morphological transition. *Adv. Colloid Interface Sci.* 96: 143.

133 Starov, V.M. (1992). Equilibrium and hysteresis contact angles. *Adv. Colloid Interface Sci.* 39: 147.

134 Hamraoui, A., Thuresson, K., Nylander, T. et al. (2000). Can a dynamic contact angle be understood in terms of a friction coefficient? *J. Colloid Interface Sci.* 226: 199.

135 Pismen, L.M., Rubinstein, B.Y., and Bazhlekov, I. (2000). Spreading of a wetting film under the action of van der Waals forces. *Phys. Fluids* 12: 480.

136 Mastrangelo, C.H. and Hsu, C.H. (1993). Mechanical stability and adhesion of microstructures under capillary forces. *J. Microelectromech. Syst.* 2: 33.

137 Mastrangelo, C.H. (1997). Adhesion-related failure mechanisms in micromechanical devices. *Tribol. Lett.* 3: 233.

138 Valli, J.A. (1986). Review of adhesion test method for thin hard coatings. *J. Vac. Sci. Technol., A* 4 (152): 3007.

139 Sheng, G., Yan, L., Liu, B. et al. (2000). Design and analysis of MEMS-based slider-suspensions for high performance magnetic recording system. *J. Micromech. Microeng.* 10: 64.

140 Zhu, Y., Liu, B., Hua, W. et al. (2000). A study of interface dynamics for stiction-free slider and super smooth disk. *J. Appl. Phys.* 87 (9): 6149.

141 Liu, B., Lim, S.T., and Sheng, G. (1997). Landing zone skew angle, long-term stiction, and contamination buildup on slider surfaces of disk drives. *J. Appl. Phys.* 81 (8): 15.

142 Liu, B., Sheng, G., and Lim, S.T. (1997). Meniscus force modeling and study on the fluctuation of stiction/friction force in CSS test process. *IEEE Trans. Magn.* 33 (5): 3121.

143 Liu, B., Chen, Q.S., Yuan, Z. et al. (1997). Tracking seeking, head-disk spacing variation and head-disk interface failure. *IEEE Trans. Magn.* 33 (5): 3136.

144 Liu, B., Chen, Q.S., Sheng, G. Skew angle and head-disk integration for Gb/in2 area density recording. The Institute of Electronics, Information and Communication Engineers (Japan) Technical Report of IEICE, MR96-39, 19, 1996.

145 Berger, E.J., Sadeghi, F., and Krousgrill, C.M. (1997). Analytical and numerical modeling of engagement of rough, permeable, grooved wet clutches. *ASME J. Tribol.* 119: 143.

146 Jang, J.Y. and Khonsari, M.M. (1999). Thermal characteristics of a wet clutch. *ASME J. Tribol.* 121 (3): 610.

147 Natsumeda, S. and Miyoshi, T. (1994). Numerical simulation of engagement of paper based wet clutch facing. *ASME J. Tribol.* 116 (2): 232.

148 Yang, Y., Lam, R.C. and Fujii, T., Prediction of Torque Response During the Engagement of Wet Friction Clutch. SAE Technical Papers, Paper number: 981097, 1998.

149 Patir, N. and Cheng, H.S. (1978). Average flow model for determining effects of three-dimensional roughness on partial hydrodynamic lubrication. *J. Lubr. Technol.* 100 (1): 12.

150 Ganemi, B., Olsson, R. and Lundström, B. Durability and Friction Characteristics of a Novel High Performance Four Wheel Drive

Automotive Transmission Fluid. SAE Technical Papers, Paper number: 2000-01-2907, 2000.

151 Mäki, R. Limited Slip Wet Clutch Transmission Fluid for AWD Differentials; Part 1: System Requirements and Evaluation Methods. SAE Technical Papers, SAE 2003-01-1980, 2003.

152 Yang, Y. Modeling of Heat Transfer and Fluid Hydrodynamics for a Multidisc Wet Clutch. SAE Technical Papers, Paper number: 950898. 1995.

153 Zagrodzki, P. (1985). Numerical analysis of temperature fields and thermal stresses in the friction discs of a multidisc wet clutch. *Wear* 101 (3): 255.

154 Mäki, R. Measurement and Characterization of Anti-Shudder Properties in Wet Clutch Applications. SAE Technical Papers, Paper number: 2005-01-0878, 2005.

155 Brecht, J., Schiffner, K. Influence of Friction Law on Brake Creep-Groan, SAE, 2001-01-3138

156 Eriksson, M. and Jacobson, S. (2001). Friction behaviour and squeal generation of disc brakes at low speeds. *Proc. Inst. Mech. Eng.* D215 (D12): 1245.

157 Eriksson, M., Lundquist, A., and Jacobson, S. (2001). A study of the influence of humidity on the friction and squeal generation of automotive brake pads. *Proc. Inst. Mech. Eng.* D215 (D3): 329.

158 Gao, H. and Barber, G.C. (2002). Microcontact model for paper-based wet friction materials. *ASME J. Tribol.* 124: 414.

159 Sheng, G., Cui, F.S. and Zhen, H., Modeling and development of computer programe for friction induced instability analysis of HGA-Disk-cassette coupling system, SONY SRL Report-MCAP/169-2, 2000.

160 Sheng, G. and Wan, Q., Brake Noise Technology, CMA Consultancy Report, Anaheim, CA, 2002.

161 Sheng, G. and Wan, Q., Automotive Friction Noise Control Technology, CMA Consultancy Report, Anaheim, CA, 2002.

162 Adams, G.G., Müftü S. and Azhar, NM. A nano scale multi-asperity contact and friction model, Proceedings of IMECE2002, ASME Congress, Nov., 2002, New Orleans, Louisiana.

163 Hurtado, J.A. and Kim, K.-S. (1999). Scale effects in friction of single asperity contacts: part I; from concurrent slip to single-dislocation-assisted slip. *Proc. R. Soc. London* A455: 3363.

164 Hurtado, J.A. and Kim, K.-S. (1999). Scale effects in friction in single asperity contacts: part II; multiple-dislocation-cooperated slip. *Proc. R. Soc. London* A455: 3385.

165 Maugis, D. (1992). Adhesion of spheres: the JKR-DMT transition using a Dugdale model. *J. Colloid Interface Sci.* 150: 243.

166 Johnson, K.L. and Greenwood, J.A. (1997). An adhesion map for the contact of elastic spheres. *J. Colloid Interface Sci.* 192: 326–333.

167 Akay, A. (2002). Acoustics of friction. *J. Acoust. Soc. Am.* 111 (4): 1525.

168 Seireg, A.A. (1998). *Friction-Induced Sound and Vibration, in Friction and Lubrication in Mechanical Design*. New York: Marcel Dekker, Chapter 11.

169 Guran, A., Pfeiffer, F., and Popp, K. (1996). *Dynamics with Friction Modeling: Analysis and Experiment*. New Jersey: World Scientific Publishing.

170 Ibrahim, R.A. (1994). Friction-induced vibration, chatter, squeal, and chaos part 1: mechanics of contact and friction. *Appl. Mech. Rev.* 47 (7): 209.

171 Ibrahim, R.A. (1994). Friction-induced vibration, chatter, squeal, and chaos part 2: dynamics and modeling. *Appl. Mech. Rev.* 47 (7): 227.

172 Nayfeh, A.H. and Mook, D.T. (1979). *Nonlinear Oscillations*. New York: Wiley.

173 Timoshenko, S., Young, D.H., and Weaver, W. (1974). *Vibration Problems in Engineering*. New York: Wiley.

174 Soedel, W. (1993). *Vibrations of Shells and Plates*, 2e. New York: Marcel Dekker, Inc.

175 Pierce, A.D. (1981). *Acoustics: an Introduction to its Physical Principles and Applications*. New York: McGraw-Hill Book Company, Inc.

176 Ingard, K.U. (1988). *Fundamentals of Waves and Oscillations*. Cambridge: Cambridge University Press.

177 Feeny, B., Guran, A., Hinrichs, N. et al. (1998). A historical review on dry friction and stick–slip phenomena. *Appl. Mech. Rev.* 51: 321.

178 Bo, L.C. and Pavelescu, D. (1982). The friction-speed relation and its influence on the critical velocity of the stick-slip motion. *Wear* 82 (3): 277.

179 Karnopp, D. (1985). Computer simulation of stick-slip friction in mechanical dynamic systems. *ASME J. Dyn. Syst. Meas. Control* 107: 100.

180 Armstrong-H'elouvry, B., Dupont, P., and Canudas de Wit, C. (1994). A survey of models, analysis tools and compensation methods for the control of machines with friction. *Automatica* 30: 1083.

181 Armstrong-H'elouvry, B. (1993). Stick-slip and control in low-speed motion. *IEEE Trans. Autom. Control* 38 (10): 1483.

182 Dupont, P.E. (1993). The effect of friction on the forward dynamics problem. *Int. J. Rob. Res.* 12 (2): 164.

183 Hess, D.P. and Soom, A. (1990). Friction at a lubricated line contact operating at oscillating sliding velocities. *J. Tribol.* 112 (1): 147.

184 Serafin, S. The sound of friction: real-time models, playability and musical applications, Thesis, Stanford university, 2004.

185 Rice, J. and Ruina, A. (1983). Stability of steady frictional slipping. *J. Appl. Mech.* 50 (2): 343.

186 Bouc, R. Forced vibration of mechanical system with hysteresis. In: Proceedings of the Fourth Conference on Nonlinear Oscillations. Prague, p. 315. 1967.

187 Wen, Y.K. (1976). Method of random vibration of hysteretic systems. *ASCE J. Eng. Mech. Div.* 102: 249.

188 Baber, T.T. and Wen, Y.K. (1981). Random vibration of hysteretic degrading systems. *ASCE J. Eng. Mech. Div.* 107: 1069.

189 Ni, Y.Q., Ko, J.M., and Wong, C.W. (1998). Identification of non-linear hysteretic isolators from periodic vibration tests. *J. Sound Vib.* 217: 737.

190 Wong, C.W., Ni, Y.Q., and Lau, S.L. (1994). Steady-state oscillation of hysteretic differential model. I. Response analysis. *ASCE J. Eng. Mech.* 120: 2271.

191 Wong, C.W., Ni, Y.Q., and Ko, J.M. (1994). Steady-state oscillation of hysteretic differential model. II. Performance analysis. *ASCE J. Eng. Mech.* 120: 2299.

192 Dahl, P.R. (1976). Solid friction damping of mechanical vibrations. *AIAA J.* 14 (12): 1675.

193 Canudas de Wit, C., Olsson, H., Aström, K.J. et al. (1995). A new model for control of systems with friction. *IEEE Trans. Autom. Control* 40 (3): 419.

194 Dupont, P. Armstrong, B., Hayward, V. Elasto-Plastic Friction Model: Contact Compliance and Stiction, 2000 ACC; Chicago, June 2000.

195 Haessig, D.A. and Friedland, B. (1991). On the modeling and simulation of friction. *J. Dyn. Syst. Meas. Contr.* 113 (3): 354.

196 Johnson, C.T. and Lorenz, R.D. (1992). Experimental identification of friction and its compensation in precise, position controlled mechanisms. *IEEE Trans. Ind. Appl.* 28 (6): 1392–1398.

197 Kim, J.-H., Chae, H.-K., Jeon, J.-Y. et al. (1996). Identification and control of systems with friction using accelerated evolutionary programming. *IEEE Control Syst. Mag.* 16 (4): 38–47.

198 Zglimbea, R., Finca, V., and Marin, C. (2008). Identification of systems with friction via distributions using the simplified Dahl model. *Int J. Math. Model Methods Appl. Sci.* 2 (2): 269–276.

199 Ravanbod-Shirazi, L. and Besançon-Voda, A. (2003). Friction identification using the Karnopp model, applied to an electropneumatic actuator. *Proc. Inst. Mech. Eng. Part I: J. Syst. Control Eng.* 217 (2): 123–138.

200 Meng, H.C. and Ludema, K.C. (1995). Wear models and predictive equations: their form and content. *Wear* 181–183: 443–457.

201 Fassois, S.D., Rizos, D., Wong, C.X. et al. (2004). Identification of pre-sliding friction dynamics. *Chaos* 14 (2): 89–95.

202 Zhang, J. Bhattacharyya, S. Simaan, N. Model and Parameter Identification of Friction during Robotic Insertion of Cochlear-Implant Electrode Arrays, 2009 IEEE International Conference on Robotics and Automation.

203 Wang, K. and Woodhouse, J. (2011). The frequency response of dynamic friction: a new view of sliding interfaces. *J. Mech. Phys. Solids* 59: 1020–1036.

204 Sheng, G. and Xu, J. (2011). A parameter identification method for thermal flying-height control sliders. *Microsyst. Technol.* 17: 1409–1415.

205 Sheng G., Sensing and identification of nonlinear dynamics of slider with clearance in sub-5 nanometer regime (Invited), *Adv. Tribol.*, Hindawi Publishing Corp., 2011, ID 282839, doi:10.1155/2011/282839.

206 Fujii, Y., Tobler W.E., and Pietron, G.M., et al. Review of Wet Friction Component Models for Automatic Transmission Shift Analysis, SAE 2003-01-1665.

207 Aleksendric, D. and Barton, D.C. (2009). Neural network prediction of disc brake performance. *Tribol. Int.* 42: 1074–1080.

208 Senatore, A., D'Agostino, V., DiGiuda, R. et al. (2011). Experimental investigation and neural network prediction of brakes and clutch material frictional behaviour considering the sliding acceleration influence. *Tribol. Int.* 44: 1199–1207.

209 Sheng, G., Huang, L., Xu, J. et al. (2012). Probing and diagnosis of slider–disk interactions in nanometer clearance regime using artificial neural network. *Microsyst. Technol.* 18 (9–10): 1255–1259.

4

Friction Dynamics of Manipulators

4.1 Friction Models of Robot Manipulator Joints

Generally, the robotic arm, an articulated mechanical structure, is an assembly of links interconnected by joints. Industry-specific robots perform several tasks such as picking, moving, and placing objects. Such robotic arms are also known as robotic manipulators.

A series of jointed links are put together to form an arm-like manipulator that is capable of automatically moving objects within a given number of degrees of freedom. Every commercial robot manipulator includes a controller and a manipulator arm. The performance of the manipulator depends on its speed, payload weight, and precision. Many manipulator joints consist of roller-bearing mechanism and transmission as well as seals and covers, in which contact and friction as well as relative motions of interfaces exist. Friction has been experimentally shown to be one of the major sources of performance degradation in motion control system. Robotic manipulators are subject to joint friction that arises in the bearings, transmissions, and seals – in other words, at every point where two surfaces are in relative motion and in contact. Numerous factors, such as surface roughness and topology, lubricant viscosity, load, contaminations, temperature, and velocity influence the friction forces at the contacting surfaces.

In a robot joint, the complex interaction of components such as gears, bearings, and shafts, which are rotating/sliding at different velocities, environmental conditions, and aging status, makes physical modeling difficult. Friction modeling is particularly critical for robot's manipulator controlling, since they can induce large positioning errors, stick–slip motions, limit cycle's instability, and chaos. It has been reported that friction can cause up to 50% error in some

Dynamics and Control of Robotic Manipulators with Contact and Friction, First Edition.
Shiping Liu and Gang (Sheng) Chen.
© 2019 John Wiley & Sons Ltd. Published 2019 by John Wiley & Sons Ltd.

industrial manipulators. Lack of or poor friction compensation action in the control may lead to significant positioning errors and system stability.

Friction is the result of complex interactions between contacting surfaces on multiscale levels. Depending on the applications, different friction models have been developed and applied in manipulator development, including the effects of speed, load, and temperatures. Empirically motivated friction models have been successfully used in many applications, including robotics. This category of models was developed through time according to empirical observations of the phenomenon. From empirical observations, it is known that friction can be affected by several factors, including temperature, force/torque levels, position, velocity, acceleration, lubricant/grease properties, wear, contamination, and aging.

Because of the demand for real applications, more work is needed to understand the influence of different factors on the friction properties. A more comprehensive friction model is needed to improve the performance of control and diagnosis of manipulator systems including friction phenomena.

Many friction models have been introduced in last chapter, including Coulomb, viscous, Stribeck effect, features of nonstationary velocities dependent, variations in the break-away force, pre-sliding behavior, small displacements occurring during the stiction phase, and hysteretic effects, etc. This gives rise to varied nonlinear expression, and combinations of them, in the form of power law, polynomial function of a proper order, or sigmoid-functions, etc. The complex model including Karnopp model, Dahl model, LuGre model. More recently, single and multistate integral friction models have been proposed, to account for the hysteresis behavior with nonlocal memory.

Many friction models have been developed for robot manipulator systems [1–44]. These models attain different levels of accuracy, and have been widely used in literature, but no one can be definitely considered more effective than the others, as many factors can significantly affect the practical implementation and the performances of each case.

For example, Leuven model is obtained by adding a hysteresis function with nonlocal memory function to the LuGre model. The modified Leuven model or Maxwell-slip friction model is very comprehensive and includes accurate modeling in the pre-sliding and sliding regimes [9]. Moreover, the extended model incorporates a hysteresis function with nonlocal memory, which forms an input–output relationship in which the output at any time depends on the previous inputs and outputs, as well as on past extremum values [13]. This offers a comprehensive integration of the generic, LuGre, and Leuven models. It is established by adding a rate-state behavior to the friction blocks of the Maxwell-slip model.

Although for the purpose of model-based friction compensation, many sophisticated friction models have been proposed in the literatures, there exists no universally agreed parametric friction model that can be used for most of real applications. This, by implication, has made the selection of an appropriate parametric friction model for specific application very difficult. The accurate determination of the parameters of these complicated parametric friction models has been very challenging due to the complexity of friction uncertainty and nonlinearities. Motivated by the need in the community for a comprehensive, nonparametric, and effective friction compensation in manipulator motion control system, an artificial intelligent (AI)-based (nonparametric) friction model has been proposed.

The recent development in AI makes it adaptable for system modeling based on data training and expert knowledge. It has been demonstrated that the major AI paradigms (neural network, fuzzy logic, support vector machine, etc.) have the capability of approximating any uncertain nonlinear functions to a reasonable degree of accuracy. These have been used to develop appropriate alternatives for friction model and compensation in manipulator motion control systems [37, 44]. In addition, the use of AI-based friction model may also reduce both the complexity and time consumed in the friction modeling and identification. For example, the friction model using support vector regression was proposed to estimate the nonlinear friction in a motion control system [44], in which it is characterized with fewer parameters for the friction model development and requires less development time. The effectiveness of the developed model in representing and compensating for the frictional effects is evaluated experimentally and the performance is benchmarked with three parametric based friction models.

The accurate modeling of the friction of manipulator joints has been the urgent need of the design and development of robot manipulator system for control purposes, where a precise friction model can substantially improve the overall performance of a manipulator in terms of accuracy and control stability. Since friction can also be determined by the status of wear and aging of joint interface and multiple scale phenomena, [19–24, 39], there is also interest in friction quantification from the perspectives of robot condition monitoring and fault detection.

LuGre friction model is a common choice for characterizing joint friction in the robotics community. Bittencourt et al. [17–19] applied the LuGre model given by Eq. (3.111) or as a generalized empirical friction model shown in Eq. (3.118) for a revolute joint:

$$\tau_f = \sigma_0 z + \sigma_1 \dot{z} + h(\dot{\varphi}_m)$$

$$\dot{z} = \dot{\varphi}_m - \sigma_0 \frac{|\dot{\varphi}_m|}{g(\dot{\varphi}_m)} z, \tag{4.1}$$

where τ_f is the friction torque and φ_m is the joint motor angle. The state z is related to the dynamic behavior of asperities in the interacting surfaces of joints and can be interpreted as their average deflection with stiffness coefficient σ_0 and damping coefficient α_1. The function $h(\dot{\varphi}_m)$ represents the velocity strengthening of viscous friction. Typically, it is given by $h(\dot{\varphi}_m) = F_v \dot{\varphi}_m$, and $g(\dot{\varphi}_m)$ quantifies the velocity weakening of friction. Motivated by the phenomena of Stribeck, $g(\dot{\varphi}_m)$ is usually modeled as

$$g(\dot{\varphi}) = F_c + F_s e^{-\left|\frac{\dot{\varphi}_m}{\dot{\varphi}_s}\right|^\alpha} \tag{4.2}$$

In which F_c is the Coulomb friction, F_s is defined as the standstill friction parameter. $\dot{\varphi}_s$ is the Stribeck velocity and α is the exponent of the Stribeck nonlinearity. This model has been successfully used to describe many of the friction characteristics [32].

Since z is not measurable, a difficulty with Eq. (4.1) is the estimation of the dynamic parameters of stiffness and damping $[\sigma_0, \sigma_1]$. In [6], these parameters are estimated in a robot joint by means of open loop experiments and by use of high-resolution encoders.

Open-loop experiments are not always possible, and it is common to apply only a static description of Eq. (4.1). For constant velocities, Eq. (4.1) is equivalent to the following static model:

$$\tau_f(\dot{\varphi}) = g(\dot{\varphi}_m) \operatorname{sign}(\dot{\varphi}_m) + h(\dot{\varphi}_m) \tag{4.3}$$

It is fully described by the g and h functions. In fact, Eq. (4.1) simply adds dynamics to Eq. (4.3). The typical choice for g and h as defined for Eq. (4.1) yields the static model

$$\tau_f(\dot{\varphi}_m) = \left[F_c + F_s e^{-\left|\frac{\dot{\varphi}_m}{\dot{\varphi}_s}\right|^\alpha} \right] \operatorname{sign}(\dot{\varphi}_m) + F_v \dot{\varphi}_m. \tag{4.4}$$

Eq. (4.4) requires a total of four parameters to comprehensively describe the velocity weakening regime $g(\dot{\varphi}_m)$ and just one parameter to capture viscous friction property $h(\dot{\varphi}_m)$.

A shortcoming of the LuGre model structure is its dependence only on the states z, which is dynamic behavior of asperities in the interface and can be interpreted as the average deflection. In more complicated and demanding applications, the effects of the other variables cannot be neglected. For instance, in [17] it was observed that a strong temperature dependence exists. In [18] it was observed that a strong joint load torque and temperature dependences exist, and the effects are considered as disturbances and estimated in an adaptive framework. However, more work is needed to fully understand the influence of different parameters and factors on the friction properties. More comprehensive friction models are needed to improve the performance and quality of the control and diagnosis of manipulator systems with friction phenomena.

4.2 Modeling Friction with Varied Effects

Due to nonsymmetry and nonuniformity in the contact surfaces, it has been observed that the friction of joint in rotating machines depends on the angular position [1–20]. It is therefore expected that this dependency is reflected in a robot joint modeling. Based on experiments, many static friction curves have been estimated in the joint angle range. Joint load torque of a manipulator can be represented by simply varying the arm configuration [17, 18]:

$$\tau_f(\dot{\varphi}_m, \tau_m) = \{F_{c,0} + F_{c,\tau_m} | \tau_m |\}$$
$$+ \{F_{s,0} + F_{s,\tau_m} | \tau_m |\} e^{-\left| \frac{\dot{\varphi}_m}{\dot{\varphi}_{s,\tau_m}} \right|^{1.3}} + F_v \dot{\varphi}_m. \quad (4.5)$$

The typical model parameters describing the dependence on τ_m are shown in Table 4.1 [17, 18].

In many applications, the friction temperature dependence is due to the change of properties of both lubricant and contacting surfaces. Based on lubrication mechanisms, both the thickness of the lubricant film and its viscosity play a critical role for the resultant friction properties. For Newtonian fluids, the shear forces of lubricant film are directly proportional to the viscosity property, and accordingly dependent on temperature [17–19]. Specific experiments have been

Table 4.1 Identified τ_m-dependent model parameters.

F_{c,τ_m}	F_s, τ_m	$\dot{\varphi}_{s,\tau_m}$
2.32×10^{-2}	1.28×10^{-1}	9.07

conducted to analyze temperature effects. In the velocity-weakening zone, a linear increase of the standstill friction with temperature has been observed. There also exist the combined effects of load torque and temperature τ_m and T on friction. For the characteristics of the temperature-related effects and load-related effects, the friction model was extended to [18, 19]:

$$\tau_f \ (\dot{\varphi}_m, \tau_m, T) = \mathcal{M}^*_{g_{\tau_m}} + \mathcal{M}^*_{g_T} + \mathcal{M}^*_{h_T} \tag{4.6a}$$

$$\mathcal{M}^*_{g_{\tau_m}} = \{F_{c,0} + F_{c,\tau_m} | \tau_m |\} + F_{s,\tau_m} \ | \ \tau_m \ | \ e^{-\left| \frac{\dot{\varphi}_m}{\dot{\varphi}_{s,\tau_m}} \right|^{1.3}}$$

$$\mathcal{M}^*_{g_T} = \{F_{s,0} + F_{s,T} T\} e^{-\left| \frac{\dot{\varphi}_m}{\{\dot{\varphi}_{s,0} + \dot{\varphi}_{s,T} T\}} \right|^{1.3}}$$

$$\mathcal{M}^*_{h_T} = \left\{ F_{v,0} + F_{v,T} \ e^{\frac{-T}{T_{V_0}}} \right\} \dot{\varphi}_m \tag{4.6b}$$

The model describes the effects of τ_m and T on friction for the investigated robot joint. According to [18, 19], the first term in Eq. (4.6b) expressions relate to the velocity-weakening friction while the third term relates to the velocity-strengthening regime. τ_m only affects the velocity weakening regime and involves in a total of three parameters, $[F_{c,\tau_m}, F_{s,\tau_m}, \dot{\varphi}_{s,\tau_m}]$. Temperature T affects both regimes and requires four parameters, $[F_{s,T}, \dot{\varphi}_{s,T}, F_{v,\tau_m}, T_{V_0}]$. The four remaining parameters, $[F_{c,0}, F_{s,0}, \dot{\varphi}_{s,0}, F_{v,0}]$, are used to relate to the original friction model structure. It is noted that in this case it is assumed that the effects of τ_m- and T are independent, their respective expressions exhibit as separated sums in the equation. Based on the already identified τ_m-dependent parameters in Table 4.1, the other parameters in the model are identified from the experimental results, after removing the τ_m-terms. The parameters are shown in Table 4.2 [17, 18].

The torque model has a total of seven terms and three parameters entering the model in a nonlinear way. The identification of such a model is important. This case illustrated that it will be important

Table 4.2 Identified T-dependent and $\mathcal{M}_0$-related model parameters.

$F_{c,0}$	$F_{c,T}$	$F_{s,0}$	$F_{s,T}$	$F_{v,0}$	$F_{v,T}$	$\dot{\varphi}_{s,0}$	$\dot{\varphi}_{s,T}$	T_{Vo}
3.04×10^{-2}	4.67×10^{-6}	-2.44×10^{-2}	1.69×10^{-3}	1.29×10^{-4}	1.31×10^{-3}	-25.00	1.00	21.00

to model, identify, and validate the torque on specific robot joints. A practical limitation of the friction model is the requirement on availability of torque and temperature. However, torque and joint temperature sensors may not be available in many standard industrial robots. For this kind of model, it is critical to have load parameters in the model to calculate the manipulation and perpendicular torques.

As a detailed example that friction is a complex phenomenon depending on many factors, the dependence on temperature and velocity is presented as follows. To quantify these combined effects, many models have been proposed in the literature. Simoni et al. [42] has selected the polynomial friction model, because it is the model that can best fit the obtained experimental friction torque data in many situations. The third-order polynomial function has been used to represent both Stribeck and Coulomb effects, and further, it has been used to represent a viscous effect that is not necessarily linear. In this regard, the friction torque function is represented as

$$\tau_f = [c_0 + c_1 |\omega| + c_2 |\omega|^2 + c_3 |\omega|^3]\mathrm{sgn}(\omega) \tag{4.7}$$

where ω is the joint velocity and it is noted that a symmetric function is considered here, in which the coefficients c_0, c_1, c_2, c_3 are the same for either positive or negative velocities. The polynomial coefficients can be determined readily by only moving one joint of the manipulator at a time. As such, each time only 1-DoF system can be treated for each joint. In the next, the modeling temperature effects is presented by following [42]. The friction torque at a given temperature is then expressed as

$$\tau_f(T) = (c_0 + c_1 |\omega| + c_2 |\omega|^2 + c_3 |\omega|^3)\,\mathrm{sgn}(\omega)[\alpha[T - T_0] + \beta] \tag{4.8}$$

The next model to be presented is a modification of the last one in which each term in the model has independent coefficients relating friction to temperature. Based on that, the relation between joint internal temperature and friction is expressed as

$$
\begin{cases}
\dfrac{dT}{dt} = [\tau_{f,RMS}\overline{\omega} - K(T - T_{env})]\dfrac{1}{C} \\
\tau_{f,RMS} = (c_0 + c_0'(T - T_0)) + (c_1 + c_1'(T - T_0)) \mid \overline{\omega} \mid + \\
(c_2 + c_2'(T - T_0)) \mid \overline{\omega} \mid^2 + (c_3 + c_3'(T - T_0)) \mid \overline{\omega} \mid^3
\end{cases} \tag{4.9}
$$

where c_0, c_1, c_2, c_3 are the coefficients representing the assumed linear relation between friction torque coefficients and temperature, and c_0', c_1', c_2', c_3' are the polynomial friction coefficients at $T = T_0$. The friction torque at a given temperature

$$
\tau_f(T) = [(c_0 + c_0'(T - T_0)) + (c_1 + c_1'\ (T - T_0)) \mid \omega \mid + \\
(c_2 + c_2'(T - T_0)) \mid \omega \mid^2 + (c_3 + c_3'(T - T_0)) \mid \omega \mid^3]\ \mathrm{sgn}(\omega). \tag{4.10}
$$

Figure 4.1 shows a schematic of a typical joint assembly, which includes a gear pair, a cycloidal transmission and a joint bearing support [38, 39]. The joint friction model can be considered as a combination of the friction models respectively associated with varied joint transmission components. Lubricated bearings and gears in the robot joints are generally accompanied by the rolling–sliding contacts in interfaces. Therefore, an appropriate friction model needs to integrate varied interfaces, including lubricated surfaces and unlubricated ones.

For the typical gear tooth pair meshing, an empirical sliding friction expression was proposed by [40, 41], based on non-Newtonian,

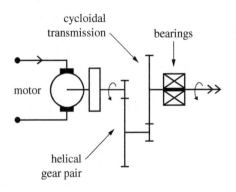

Figure 4.1 Schematic of a typical joint assembly [38, 39].

thermal elastohydrodynamic lubrication (EHL) theory. It is given by

$$\mu_{Xi}(t) = e^{f(SR_i(t), P_{hi}(t), \eta_M, S_{avg})} P_{hi}^{b_2} \mid SR_t(t) \mid^{b_3} \quad V_{ei}^{b_6}(t) \eta_M^{b_1} R_i^{b_8}(t)$$
$$\times \text{sgn}\,[\text{mod}(\Omega_p r_{bp} t, \lambda) + (n - i)\lambda - L_{AP}],$$
$$f(SR_i(t), P_{hi}(t), \eta_M, S_{avg}) = b_1 + b_4 \mid SR_i(t) \mid P_{hi}(t)\log_{10}(\eta_M)$$
$$+ b_5 e^{-|SR_i(t)|P_{hi}(t)\log_{10}(\eta_M)} + b_9 e^{S_{avg}},$$

(4.11)

Xu et al. [40, 41] suggested the following empirical coefficients (in consistent units) for the above formula:
$b_1 = -8.916465, b_2 = 1.03303, b_3 = 1.036\,077, b_4 = -0.354\,068,$
$b_5 = 2.812\,084, b_6 = -0.100\,601, b_7 = 0.752755, b_8 = -0.390\,958,$
$b_9 = 0.620\,305.$

4.3 The Motion Equations of Dynamics of Robot Manipulators with Friction

4.3.1 The General Motion Equation of Robot Manipulators

A fast and high-fidelity dynamic model is very useful for the planning, control, and estimation of robot manipulators. Obtaining the dynamic model of a robot manipulator with friction plays an important role in the description, design, and development of the system behavior for specific requirements and specifications. This requires describing the system motion, analyzing configurations, and designing control strategy for a specific manipulator. A robot manipulator is basically a positioning and handling device. To control the position we must know the dynamic properties of a manipulator so as to determine how much force must be applied on it to cause the effects. There are many reasons for studying the dynamics of a manipulator, including simulation, analysis, and synthesis of suitable control strategy, analysis of the system properties of the manipulator, testing desired motions, and verification. Accurately deriving the dynamic equation of motion for robot manipulators is difficult due to the large number of degrees of freedom, uncertainty, and nonlinearities presented in the system. There are two methods for deriving the dynamic model equation of a robot manipulator. The first method is based on the Euler–Lagrange formulation and the second one is Euler–Lagrange's methodology

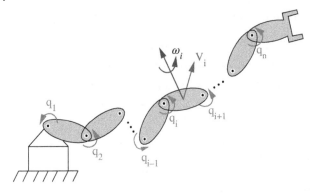

Figure 4.2 Schematic of multiple link manipulator [51].

[45–65]. In the next, we present the Euler–Lagrange method, which helps to provide good insight into controller design related to state space and provides a closed-form interpretation of the various components in the dynamic model, including inertia, gravitational effects, friction (joint/link/transmission/motor driver), Coriolis forces relating motion of one link to coupling effects of other links' motions, centrifugal forces that cause the link to have a tendency to "fly away," as well as the coupling to neighboring links and its own motion.

To derive the motion equation using Euler–Lagrange model, consider a multiple link manipulator as shown in the Figure 4.2.

Direct model equation can be derived once the forces/torques is applied to the joints, as well as the joint positions and velocities are known, compute the joint accelerations.

Generally, a manipulator consists of an open kinematic chain, whose dynamic model is affected by several factors including low rigidity or elasticity in the structure and in the joints, dynamic coupling among links, as well as potentially unknown parameters in operations (dimensions, inertia, mass). Other nonlinear effects are usually introduced by the actuation system such as *dead zones as well as friction. In the following, in the derivation of the dynamic model, an ideal case of* a series of connected rigid bodies is assumed. The dynamic model equation can be used for computation of the time history of displacement, velocity, and acceleration for the given generalized force/torque inputs applied to the joints by motor/actuator, and for known external forces applied to the end-effector, and for the given initial conditions. We assume that for the manipulator system, the kinetic energy function $K(q, \dot{q})$ and the potential energy function $P(q)$, and therefore the Lagrange function and dynamics equations can be derived.

The kinetic energy of a manipulator can be determined for each link, for given link mass, moment inertia of mass, linear velocity, and the rotational velocity the link. The kinetic energy K_i of the i-th link has the form

$$K_i = \frac{1}{2}m_i \mathbf{v}_{Ci}^T \mathbf{v}_{Ci} + \frac{1}{2}\omega_i^T \mathbf{R}_i \widetilde{\mathbf{I}}_i \mathbf{R}_i^T \omega_i \tag{4.12}$$

In which $\mathbf{R}_i$ is the rotation matrix between the frame fixed to the link and mass center. Next it is necessary to compute the linear and rotational velocities (v_{Ci} and ω_i) as functions of the Lagrange coordinates in which the joint variables are q.

The end-effector velocity may be computed as a function of the joint velocities $\dot{q}_1, \cdots$ through the Jacobian matrix $\mathbf{J}$. The same method can be used to compute the velocity of a generic point on manipulator, and in particular the velocity center of mass, which results in a function of the joint velocities only:

$$\dot{\mathbf{P}}_{Ci} = \mathbf{J}_{v1}^i \dot{q}_1 + \mathbf{J}_{v2}^i \dot{q}_2 + \ldots + \mathbf{J}_{vi}^i \dot{q}_i = \mathbf{J}_v^i \dot{\mathbf{q}}$$

$$\omega_i = \mathbf{J}_{\omega 1}^i \dot{q}_1 + \mathbf{J}_{\omega 2}^i \dot{q}_2 + \ldots + \mathbf{J}_{\omega i}^i \dot{q}_i = \mathbf{J}_\omega^i \dot{\mathbf{q}}$$

where

$$\mathbf{J}_v^i = [\mathbf{J}_{v1}^i \quad \cdots \quad \mathbf{J}_{vi}^i \quad 0 \quad \cdots \quad 0]$$

$$\mathbf{J}_\omega^i = [\mathbf{J}_{\omega 1}^i \quad \cdots \quad \mathbf{J}_{\omega i}^i \quad 0 \quad \cdots \quad 0] \tag{4.13}$$

with

$$\begin{bmatrix} \mathbf{J}_{vj}^i \\ \mathbf{J}_{\omega j}^i \end{bmatrix} = \begin{bmatrix} \mathbf{z}_{j-1} \times (\mathbf{p}_{Ci} - \mathbf{p}_{j-1}) \\ \mathbf{z}_{j-1} \end{bmatrix}$$

$$\begin{bmatrix} \mathbf{J}_{vj}^i \\ \mathbf{J}_{\omega j}^i \end{bmatrix} = \begin{bmatrix} \mathbf{z}_{j-1} \\ 0 \end{bmatrix}$$

in which $\mathbf{p}_{j-1}$ is the position of the origin of the frame associated to the j-th link. For a n-DoF manipulator, we have:

$$K = \frac{1}{2}\sum_{i=1}^{n} m_i \mathbf{v}_{Ci} + \frac{1}{2}\sum_{i=1}^{n} \omega_i^T \mathbf{R}_i \widetilde{\mathbf{I}}_i \mathbf{R}_i^T \omega$$

$$= \frac{1}{2}\dot{\mathbf{q}}^T \sum_{i=1}^{n} [m_i \mathbf{J}_v^{i\,T}(\mathbf{q}) \mathbf{J}_v^i(\mathbf{q}) + \mathbf{J}_\omega^{i\,T}(\mathbf{q}) \mathbf{R}_i \widetilde{\mathbf{I}}_i \mathbf{R}_i^T \mathbf{J}_\omega^i(\mathbf{q})] \, \dot{\mathbf{q}}$$

$$= \frac{1}{2}\dot{\mathbf{q}}^T \mathbf{M}(\mathbf{q})\dot{\mathbf{q}}$$

$$= \frac{1}{2}\sum_{i=1}^{n}\sum_{j=1}^{n}Mi_j(\mathbf{q})\dot{q}_i\dot{q}_j \tag{4.14}$$

In which $\mathbf{M}(\mathbf{q})$ is a $n \times n$, symmetric and positive definite matrix, function of the manipulator configuration q. Matrix $\mathbf{M}(\mathbf{q})$ is the so-called the inertia matrix of the manipulator.

Computation of the potential energy is given as follows. For rigid bodies, the only potential energy taken into account in the dynamics is due to the gravitational field **g**. For the generic i-th link,

$$P_i = \int_{L_i} \mathbf{g}^T \mathbf{p}\, dm = \mathbf{g}^T \int_{L_i} \mathbf{p}\, dm \; \mathbf{g}^T \mathbf{p}_{Ci} m_i \tag{4.15}$$

The potential energy does not depend on the joint velocities $\dot{q}$, and may be expressed as a function of the position of the centers of mass. For the whole manipulator system,

$$P = \sum_{i=1}^{n} \mathbf{g}^T \mathbf{p}_{Ci} m_i \tag{4.16}$$

In case of flexible link, one should consider also terms due to elastic forces.

P is computed using a procedure similar to the one for the kinematic energy K. Once K and P are known, it is possible to compute the dynamic model of the manipulator. The dynamics is expressed as,

$$\psi_k = \frac{d}{dt}\left(\frac{\partial \mathcal{L}}{\partial \dot{q}_k}\right) - \frac{\partial \mathcal{L}}{\partial q_k} \qquad k = 1, \dots, n \tag{4.17}$$

The Lagrange function is given by

$$\mathcal{L} = K - P = \frac{1}{2}\sum_{i=1}^{n}\sum_{j=1}^{n}M_{ij}\dot{q}_i\dot{q}_j - \sum_{i=1}^{n}\mathbf{g}^T \mathbf{p}_{Ci} m_i \tag{4.18}$$

Then

$$\frac{\partial \mathcal{L}}{\partial \dot{q}_k} = \frac{\partial K}{\partial \dot{q}_k} = \sum_{j=1}^{n}M_{kj}\dot{q}_j \tag{4.19a}$$

$$\frac{d}{dt}\frac{\partial \mathcal{L}}{\partial \dot{q}_k} = \sum_{j=1}^{n}M_{kj}\ddot{q}_j + \sum_{j=1}^{n}\frac{d\,M_{kj}}{dt}\dot{q}_j = \sum_{j=1}^{n}M_{kj}\ddot{q}_j + \sum_{i=1}^{n}\sum_{j=1}^{n}\frac{\partial}{\partial q_i}\dot{q}_i\dot{q}_j \tag{4.19b}$$

Moreover,

$$\frac{\partial \mathcal{L}}{\partial q_k} = \frac{1}{2} \sum_{i-1}^{n} \sum_{i-1}^{n} \frac{\partial M_{ij}}{\partial q_k} \dot{q}_i \dot{q}_j - \frac{\partial P}{\partial q_k} \tag{4.19c}$$

The Lagrange equations have the following formulation:

$$\sum_{j=1}^{n} M_{kj} \ddot{q}_j + \sum_{i=1}^{n} \sum_{j=1}^{n} \left[\frac{\partial M_{kj}}{\partial q_k} - \frac{1}{2} \frac{\partial M_{ij}}{\partial q_k} \right] \dot{q}_i \dot{q}_j + \frac{\partial P}{\partial q_k} = \psi_k$$

$$k = 1, \dots, n \tag{4.20}$$

We can define the term $h_{kji}(\mathbf{q})$ as

$$h_{kji}(\mathbf{q}) = \frac{\partial M_{kj}(\mathbf{q})}{\partial q_i} - \frac{1}{2} \frac{\partial M_{ij}(\mathbf{q})}{\partial q_k} \tag{4.21}$$

and $g_k(\mathbf{q})$ as

$$g_k(\mathbf{q}) = \frac{\partial P(\mathbf{q})}{\partial q_k}$$

Then the following equations are finally obtained:

$$\sum_{i=1}^{n} M_{kj}(\mathbf{q}) \; \ddot{q}_j + \sum_{i=1}^{n} \sum_{j=1}^{n} h_{kji}(\mathbf{q}) \; \dot{q}_i \dot{q}_j + g_k(\mathbf{q}) = \psi_k \quad k = 1, \dots, n$$

$$\tag{4.22}$$

The elements $M_{kj}(\mathbf{q})$, $h_{kji}(\mathbf{q})$, $g_k(\mathbf{q})$ are function of the joint position only, and therefore their computation is relatively simple once the manipulator's configuration is known. They have the following physical meaning. For the acceleration terms: $M_{kk}(\mathbf{q})$ is the moment of inertia about the k-th joint axis, in a given configuration and considering blocked all the other joints. $M_{kj}(\mathbf{q})$ is the *inertia coupling*, accounting for the effect of acceleration of joint j on joint k.

For the quadratic velocity terms: $h_{kji}(\mathbf{q})(dq/dt)^2$ represents the centrifugal effect induced on joint k by the velocity of joint j. (notice $h_{kkk} = \partial M_{kk}/\partial q_k = 0$). $h_{ijk}(\mathbf{q}) \dot{q}_i \dot{q}_j$ represents the Coriolis effect induced on joint k by the velocities of joints i and j.

For the configuration-dependent terms, g_k represents the torque generated on joint k by the gravity force acting on the manipulator in the current configuration.

It is noted that the nonconservative forces ψ_k are, in general, composed of joint actuator torque; external (contact) forces, and joint friction torque.

The Lagrangian equations

$$\sum_{j=1}^{n} M_{kj}(\mathbf{q}) \ \ddot{q}_j + \sum_{i=1}^{n} \sum_{j=1}^{n} h_{kji}(\mathbf{q}) \ \dot{q}_i \dot{q}_j + g_k(\mathbf{q}) = \psi_k \quad k = 1, \dots, n$$

(4.23a)

can be written in matrix form as

$$\mathbf{M}(\mathbf{q})\ddot{\mathbf{q}} + \mathbf{C}(\mathbf{q}, \dot{\mathbf{q}})\dot{\mathbf{q}} + \mathbf{D}\dot{\mathbf{q}} + \mathbf{g}(\mathbf{q}) = \tau + \mathbf{J}^{\mathrm{T}}(\mathbf{q})\mathbf{F}_c$$

(4.23b)

The above multivariable, nonlinear system is known as the dynamic model of the manipulator.

$$\mathbf{M}(\mathbf{q})\ddot{\mathbf{q}} + \mathbf{C}(\mathbf{q}, \dot{\mathbf{q}})\dot{\mathbf{q}} + \mathbf{D}\dot{\mathbf{q}} + \mathbf{g}(\mathbf{q}) + \tau_f = \tau$$

(4.24)

where $\mathbf{q}$ is the vector of robot angles at arm, $\mathbf{M}$ is the inertia matrix, $\mathbf{C}$ relates to speed-dependent terms (e.g. Coriolis and centrifugal), $\mathbf{g}$ are the gravity-induced torques and τ_f contain the joint friction components. The system is controlled through the input torque, τ applied to the joint motor.

In deriving the above dynamic model, the actuation system has not been taken into account. This is normally composed of motors and reduction gears transmission system. The actuation system has several effects on the dynamics: if motors are installed on the links, then masses and inertia are changed, it introduces its own dynamics (electromechanical, pneumatic, hydraulic, etc.) that must be taken into account. It may introduce additional nonlinear effects such as backslash, friction, elasticity, etc. It is noted that these varied effects could be considered by introducing suitable terms in the dynamic model derived on the basis of the Lagrange formulation.

The friction models are presented in Section 4.2. The basic friction model may consist of lubricated roller bearings of joints with rolling–sliding contact modeled by viscous friction and friction caused by asperity contacts as well as the contribution from gear transmission in some manipulators. The motor torque inputs are briefed as follows. The motor model we are considering here is an ideal DC motor for most robotics actuators. An idealized model for an ideal DC motor can be derived by considering the power balance present in the motor at a constant voltage U,

$$P_{el} = P_{mech} + P_J,$$

(4.25)

where P_{el} is the input electrical power, P_{mech} denotes the mechanical output power, and P_J is due to the Joule heating losses in the motor operating. Eq. (4.25) can be reconfigured as

$$UI = \omega\tau + RI^2$$

where U is the voltage applied to the motor, I is the current owing through the winding, ω and τ are respectively the angular velocity and torque, and R denotes the motor winding resistance. The torque constant k_τ describes the relationship between t and I:

$$\tau = k_\tau I,$$

which yields

$$U = \omega k_\tau + \frac{R}{k_\tau}\tau. \tag{4.26}$$

This equation specified the required motor voltage to produce a target torque for a given angular velocity. For a manipulator system, these can be calculated using the inverse dynamics.

The external forces can be divided into two components. Let τ_i represent the motor applied torque at the joint and define N_i to be any other forces that act on the i-th generalized coordinate, including conservative forces arising from a potential as well as frictional forces. The negative sign in the definition of N_i is apparent in a detailed moment. As an example, if the manipulator has viscous friction at the joints, then N_i would be a function of velocity. Other forces acting on the manipulator, such as forces applied at the end-effector, can also be included by inserting them to the joints through the transpose of the appropriate terms.

In deriving the above dynamic model Eqs. (4.23a) and (4.23b), the flexibility of system has not been taken into account. This normally needs to treat links to be flexible element instead of rigid bodies [66–73]. For example, some research introduces a friction observer for robots with joint torque sensing, such as, in particular, for the DLR medical robot, so as to increase the positioning accuracy and the performance of torque control. The observer output specifies the low-pass filtered friction torque. It can be used for friction compensation in conjunction with a specific controller designed for flexible joint arms. Some passivity analysis has been done for this friction compensation, allowing a Lyapunov-based convergence analysis in the context of the nonlinear robot dynamics. For the completely controlled system, global asymptotic stability needs to be shown, and experimental results are needed to validate the practical efficiency of the approach.

For more complicated cases, the fractional calculus model is even used for the modeling and dynamic simulations of nonlinear robotic manipulator [74]. The Riemann-Liouville approach and the fractional Euler–Lagrange equations are used to obtain the fractional-order nonlinear dynamics equations of a two-link robotic manipulator. The equations simulated for several cases involves in the integer and noninteger order analysis, with and without external forcing acting and some different initial conditions. The fractional nonlinear governing equations of motion are coupled and the time history of the angular positions and the phase diagrams can be used to visualize the effect of fractional order approach. The recent results reveal that the fractional-nonlinear robotic manipulator can exhibit different and curious behavior from those obtained with the standard dynamical system and can be useful for a better understanding and control of such nonlinear manipulator systems.

4.3.2   The Motion Equation of Two-Link Robot Manipulators

Next, we present a two-link robotic manipulator model as an example. Figure 4.3 shows the schematic.

The two-link manipulator consists of two rigid links and two rotary or revolute joints, with the first one attached to the ground

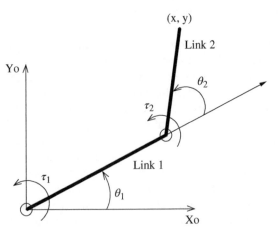

Figure 4.3
Schematic of a typical two-link manipulator.

through a joint and attached to the second link through the second joint. Motors and sensors are installed at the joints to drive the links through desired angles and for feedback control. We can position the tip of the second link at any arbitrary Cartesian position inside its workspace. The relationships between the joint variable (θ_1, θ_2), and the Cartesian variables, (x, y) can be easily obtained. The task of the manipulator to be performed by the manipulator could involve moving the end of the second link from one point to another or moving the end along a desired trajectory for trajectory tracking. The control strategy of the manipulator is to determine the time history of motor torques required to allow link to execute the desired motion. Many control strategies are proposed for this purpose. One of the conventional algorithms is the independent joint proportional derivative controller (PD), in which the motor torque is proportional to the error (difference between desired joint rotation and the measured joint rotation), and to the ratio of the change of error. Some control strategy uses the dynamical model together with the error driven portion [38–65]. These two algorithms yield stable manipulator operation, for point-to-point and trajectory tracking, under certain conditions on the controller gains and estimates of the model. These algorithms could be robust to changers in controller gains and model parameters [38–65].

However, the nonlinear ordinary differential equations of the manipulator with PD or model-based control can exhibit chaos for certain ranges of controller gains and for large mismatch in model parameters. For the manipulator model shown in Figure 4.3, the manipulator has two links with length l_1 and l_2. The dynamic equation can be derived as

$$[m_1 r_1^2 + I_1 + I_2 + m_2 r_2^2 + m_2 l_1^2 + 2m_2 l_1 r_2 \cos(\theta_2)]\theta_1$$
$$+ [m_2 r_2^2 + I_2 + m_2 l_1 r_2 \cos(\theta_2)]\ddot{\theta}_2$$
$$- m_2 l_1 r_2 \sin(\theta_2)[2\dot{\theta}_1 + \dot{\theta}_2]\dot{\theta}_2 + \tau_{1f} = \tau_1$$
$$[m_2 r_2^2 + I_2 + m_2 l_1 r_2 \cos(\theta_2)]\ddot{\theta}_1 + [m_2 r_2^2 + I_2]\ddot{\theta}_2$$
$$+ m_2 l_1 r_2 \sin(\theta_2)\dot{\theta}_1^2 + \tau_{2f} = \tau_2. \tag{4.27}$$

in which m_i, l_i, I_i, r_i are the mass, length, inertia, and position of the center of mass of link, respectively, τ_1, τ_2 are the actuating torques at the two joints, and τ_{1f}, τ_{2f} are the friction torques at the two joints.

Consider that each joint-instrumented sensor has the joint angle measured, and the motor at each joint applies a desired torque on the neighboring link. To realize point-to-point or trajectory tracking, the motor must be commanded in a continuous manner, necessitating the use of some kind of control system to compute the appropriate motor commands that will realize a desired motion. The independent joint control strategy is defined as the following computed joint torque [38–65]:

$$\Gamma = \theta_d + K_v E + K_p \dot{E} \tag{4.28}$$

in which $E = \theta_d - \theta$ is the servo error and $\dot{E}$ is the derivative of the servo error $\Gamma = (\tau_1, \tau_2)$.

This scheme is the proportional and derivative (PD) control scheme in which a term proportional to the integral of the error is added for reducing steady-state errors. This scheme was proved to be asymptotically stable for any positive velocity gain ($K_v > 0$) for a regulator or set point tracking problem ($\dot{\theta}_d = \ddot{\theta}_d = 0$). However, such results do not exist for the robot joint trajectory tracking problem ($\dot{\theta}_d \neq 0$ and $\ddot{\theta}_d \neq 0$).

In this control strategy, each joint is controlled as a separate control system without considering any of the dynamic coupling between joints. However, a model-based control strategy incorporates a complete dynamical model of the system into the computation of the actuating torques.

In the past decade, many control strategies have been explored to obtain ideal certain motions of a two-link manipulator. Recent studies have confirmed that the open-plus-close-loop (OPCL) control strategy is powerful for complicated dynamic systems of manipulators.

4.4 Nonlinear Dynamics and Chaos of Manipulators

Generally, the ordinary differential equation of robot manipulator shown in Eqs. (4.23a) and (4.23b) is a nonlinear stochastic dynamics equation, which could exhibit very rich complicated properties of nonlinear dynamics and stochastic dynamics due to the uncertainty and nonlinearity of frictions. For example, the manipulator having above PD feedback controller has been demonstrated to have varied nonlinear properties including chaotic motions.

Nonlinear stochastic dynamic systems can display complex behaviors, such as the following;

- Multiple steady-state periodic solutions are sometimes stable and sometimes unstable, in response to the same inputs.

- Jump phenomena involve discontinuous and significant changes in the response of the system as some forcing parameters are slowly varied.

- Response is at frequencies other than the forcing frequency internal resonances, involving different parts of the system vibrating at different frequencies, all with steady amplitudes (the frequencies are usually in rational ratios, such as 1/2, 1/3, 3/5, etc.).

- Self-sustained oscillations occur in the absence of explicit external periodic forcing.

- Complex, irregular motions that are extremely sensitive to initial conditions and system parameters (chaos); the system property evolves radically with the change in system parameters (bifurcation).

- System parameters and solutions exhibit statistical property and have uncertainty characteristics.

For varied cases of robot manipulators, many researches exist on bifurcations and chaos, referring only to numerical or experimental results [75–98]. Some research numerically demonstrated chaos as a motion characterized by sensitive dependence on initial conditions and topological transitivity. Also, they propose some controllers to accomplish several control objectives, showing some dynamical properties from numerical experiments. Mahout et al. [75] presented simulations of a complex behaviors of a robot manipulator with complicated nonlinear dynamics properties: harmonic, sub-harmonic, higher harmonic, fractional harmonic, and chaotic responses. Ravishankar and Ghosal [77–79] analyzed a controlled 2-DoF robot manipulator. Control of mechanical manipulators is a field of intense activity in robotics [38–65]. Among the procedures proposed to regulate mechanical manipulators, the proportional-derivative (PD) compensator is widely used. Under some circumstances, a complex dynamical behavior may arise in this system. In particular, the PD-controlled pendulum may exhibit chaotic behavior when the reference for the angular position is periodic and the total dissipation and proportional gain satisfy some conditions. This fact has been formally shown by using Melnikov's theory for detecting transversal intersections of invariant manifolds arising from a perturbed homoclinic trajectory of the nominal system. This method is one of the few analytical techniques to predict homoclinic chaos, which is one of the many forms of strange behavior.

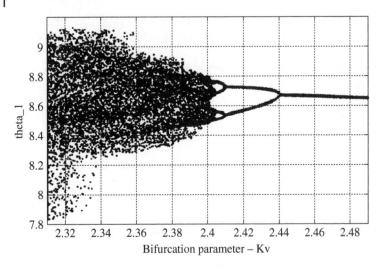

Figure 4.4 Bifurcation diagram for $\theta 1$ vs. Kv (PD controller; Kp = 52) [77, 78].

Ravishankar and Ghosal [77–79] demonstrated that the nonlinear ordinary differential equation of Eq. (4.24), which describe the motion of a feedback-controlled two-revolute robot (undertaking repetitive motions), could display chaos. They presented plots of bifurcation diagrams and used Lyapunov exponents for testing chaos, as shown in Figures 4.4 [77, 78].

Mahout et al. [75] proved that a two-revolute joint manipulator controlled with proportional- derivative (PD) law exhibited chaotic dynamics for certain values of static variables. Ravishankar and Ghosal [77–79] probed chaos in feedback-controlled 2- and 3-DoF robots. Nonlinear ODEs, which depicted the dynamics of a feedback-controlled rigid robot, demonstrated chaos for a certain range of parameters. These analyses signify that chaos is intrinsic to robot dynamics.

It is known that the performance of such a mechanism is influenced by system parameters such as mass and friction coefficient, the initial states, and external input such as driving torque controlled by specially designed controller. Conversely, study on these motions can provide a novel way to improve system's performance, optimize structure design, and develop new control strategy. For a two-link manipulator, the motions of its two rotating links can be single periodic, multiple periodic, quasi-periodic, and chaotic.

In addition to the PD feedback controller to obtain chaotic motions for a two-link manipulator as in last section, the other bifurcation characteristics of motions of a two-link manipulator were presented in [80–90].

The motions of single periodic, multiple periodic, quasi-periodic, and chaotic of a two-link manipulator with an OPCL controller have also been studied [95–97]. The dynamics of a two-link manipulator is modeled with nonlinear differential equations, followed by the Lyapunov stability analysis of the controlled system.

In motion Eq. (4.27), τ_{1f}, τ_{2f} are friction forces coming from the two joints of the two-link manipulator, and they are stated in following section. m_i is the mass of link i, I_i is the moment of inertia of link i with respect to its mass center, d_i is the distance between link i and the joint i; **g** is the acceleration of gravity; i is driving moment on the joint i, $i = 1, 2$.

Only considering the viscous friction, the joint frictions of joint 1 and joint 2, namely, f_1 and f_2 in Eq. (4.24), are simplified as follows [97].

$$f_1 = f_v \dot{\theta}_1, \qquad f_2 = f_v \dot{\theta}_2 \tag{4.29}$$

where, f_v is the viscous friction coefficient. The OPCL controller is integrated to the system for joint torques [96, 97]. They demonstrated the existence of motion bifurcation along the viscous frictions. Here only the effects of the joint viscous frictions on the joint angle motions are discussed. The goal trajectories of the two joints of the manipulator are designed as sinusoidal functions. For specified control parameters, if the joint frictions are completely ignored, namely, $f_1 = 0, f_2 = 0$, the manipulator can achieve synchronous motions.

When different values of the viscous joint friction coefficient f_v are applied, we can obtain some different types of motions such as single periodic, multiple periodic, quasi-periodic, and chaotic motions of the two joints. The motion bifurcation diagrams for the two rotating joints are shown in Figure 4.5 using the shoot method.

The bifurcation diagrams are along the bifurcation parameter of f_v changing from 0 to 4.15.

From Figure 4.5, it can be seen that θ_1 and θ_2 display the same bifurcation characteristics obviously. When the value of f_v changes between 0 and 2.1, θ_1 and θ_2 have the periodical characteristic. The jump phenomenon happens at $f_v = 1.94$. When the value of f_v changes between 2.1 and 2.7, θ_1 and θ_2 appear chaotic characteristics together with periodic motion windows. When the value of f_v changes between 2.7 and 3.7, θ_1 and θ_2 are periodic, too. When the value of f_v changes

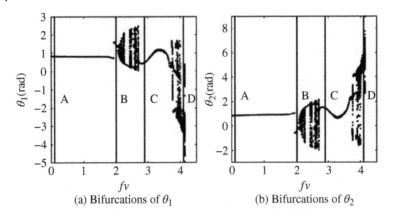

(a) Bifurcations of θ_1 (b) Bifurcations of θ_2

Figure 4.5 Motion bifurcations along viscous joint friction f_v [97].

between 3.7 and 4.15, θ_1 and θ_2 appear chaotic characteristics again, together with periodic motion windows.

There are four different cases for the value of f_v, chosen as 0.1, 2.01, 2.9, and 4.1, respectively, with respect to the four straight lines noted as A, B, C, and D in Figure 4.5, respectively. These four motions are listed in the Table 4.3 and demonstrated in the following sections.

The multiple periodic motions can be achieved when the viscous friction coefficient $f_v = 2.04$ as shown in Figure 4.6.

It can be seen that the motions of θ_1 and θ_2 are also periodical but with other harmonic components in Figure 4.6a,b). There are six frequency lines in the corresponding amplitude spectra of θ_1 and θ_2, respectively, as shown in Figure 4.6c,d). Accordingly, both of the

Table 4.3 The different motions of the two-bar linkage with different friction values.

Case no.		Motion patterns	Poincare mapping	Amplitude spectra
A	0.1	Single-periodic	One isolated point	Single frequency line
B	2.01	Multiple-periodic	Isolated points	Several obvious frequency lines
C	2.9	Quasi-periodic	Straight line with isolated points	Several obvious frequency lines
D	4.1	Chaotic	Regular shape with unlimited points	Limited broadband spectra

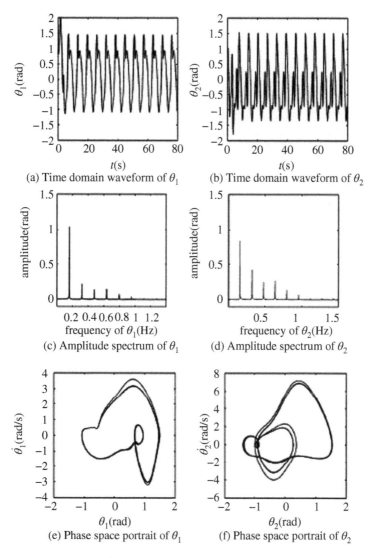

Figure 4.6 The simulated multiple periodic motions: (a and b) time-domain waveform; (c and d) amplitude spectrum; (e and f) phase space portrait; (g and h) Poincare section [97].

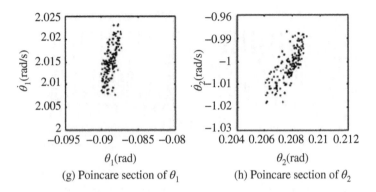

(g) Poincare section of θ_1 (h) Poincare section of θ_2

Figure 4.6 (*Continued*)

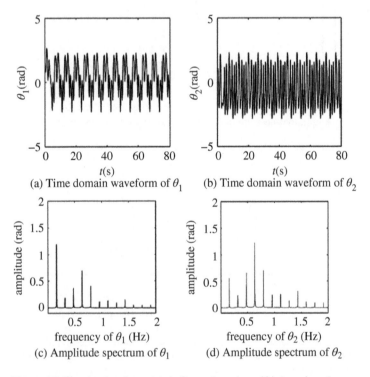

(a) Time domain waveform of θ_1 (b) Time domain waveform of θ_2

(c) Amplitude spectrum of θ_1 (d) Amplitude spectrum of θ_2

Figure 4.7 The simulated quasiperiodic motions: (a and b) time-domain waveform (c and d) amplitude spectrum; (e and f); phase space portrait; (g and h) Poincare section [97].

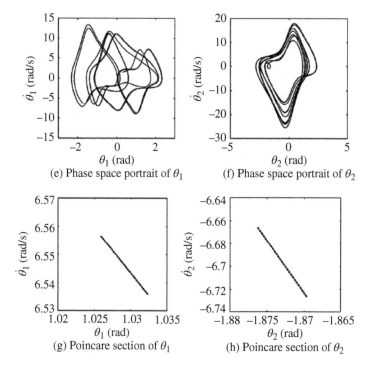

(e) Phase space portrait of θ_1

(f) Phase space portrait of θ_2

(g) Poincare section of θ_1

(h) Poincare section of θ_2

Figure 4.7 *(Continued)*

phase space portraits, as shown in Figure 4.6e,f), are closed curves, and the Poincare mapping portraits, shown in Figure 4.6g,h), give some isolated points. These indicate that the multiperiodic motion of the manipulator moves stably in this case.

The manipulator can also achieve the quasiperiodic motions. The simulation results with respect to the viscous friction coefficient $f_v = 2.9$ are shown in Figure 4.7. From Figure 4.7a,b), it can be seen that the motions of θ_1 and θ_2 are complex but regular. There are 9 and 12 obvious frequency lines for θ_1 and θ_2, respectively, shown in their corresponding amplitude spectra in Figure 4.7c,d). Both of the two-phase space portraits, as shown in Figure 4.7e,f), are difficult to distinguish from traditional harmonics, and the Poincare mapping portraits of them, shown in Figure 4.7g,h), are the concentrated sticklike areas. These can help to conclude that the simulated motions of the manipulator are quasiperiodic in this case.

The manipulator also presents chaotic motions when the viscous friction coefficient $f_v = 4.1$ shown in Figure 4.8. The simulated responses of θ_1 and θ_2 are irregular without obvious periods, as shown in Figure 4.8a,b). Their corresponding amplitude spectra as shown in Figure 4.8c,d), demonstrate multifrequency lines and definite broadband ranges. Moreover, both of the phase space portraits (shown in Figure 4.8e,f)) and the Poincare mapping (shown in Figure 4.8g,h)) of them illustrate irregular shape. According to the qualitative theory of chaos, these motions exhibit chaotic behaviors. In addition, the obtained chaotic signals shown in Figure 4.5 are evaluated via nonlinear analysis methods. The first five-order Lyapunov exponents of the time series of θ_1 and θ_2 are calculated and shown in Table 4.4. Among them, the maximum Lyapunov exponents for θ_1 and θ_2 are positive, i.e. 0.1432 and 0.1742, respectively; and the sum of the

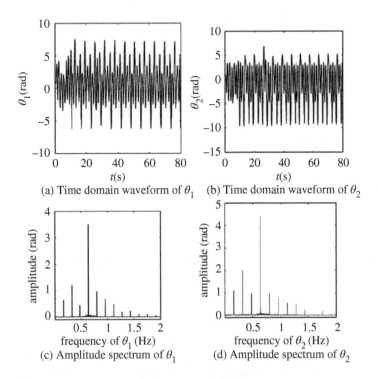

(a) Time domain waveform of θ_1 (b) Time domain waveform of θ_2

(c) Amplitude spectrum of θ_1 (d) Amplitude spectrum of θ_2

Figure 4.8 The simulated chaotic motions: (a and b) time-domain waveform; (c and d) amplitude spectrum; (e and f) phase space portrait; (g and h) Poincare section [97].

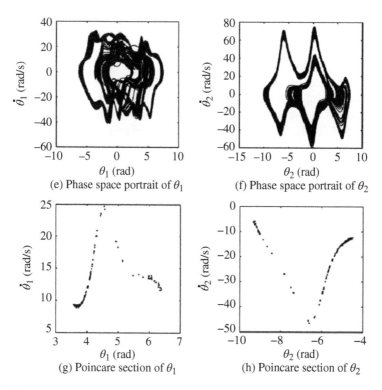

(e) Phase space portrait of θ_1

(f) Phase space portrait of θ_2

(g) Poincare section of θ_1

(h) Poincare section of θ_2

Figure 4.8 (*Continued*)

rest Lyapunov exponents are both smaller than zero. The calculated hypothesis possibilities of them with the surrogate data method are 4.2526×10^{-14} and 5.9422×10^{-15}, respectively. The two possibility values are both greatly smaller than 0.05, an empirical given threshold for chaos. Therefore, it can be seen that the simulated joint rotating angles of θ_1 and θ_2 in Case D as shown in Figure 4.8 are chaotic.

Table 4.4 Lyapunov exponents of the simulated chaos motions.

Signal	First order	Second order	Third order	Forth order	Fifth order
θ_1	0.1432	−0.0375	−0.0789	−0.1566	−2.4472
θ_2	0.1742	−0.0287	−0.0708	−0.1509	−2.5306

4.5   Parameters Identification

4.5.1   Identification of Dynamic Parameters

Robotic manipulator applications require an accurate model to perform tasks where dynamics is significant. The friction model discussed in Section 4.2 aims at improving the existing rigid robot model. The losses in joint transmission originate in friction between moving parts in contact or between moving parts and the ambient fluid. Commonly, robotic identification models represent joint transmission friction force as a viscous type one, depending on velocity, added to a constant dry friction force. However, the tribology science suggests that friction in general depends on multiple factors including load (contact force) and temperatures, etc. It is important to consider this complicated dependence when variable loads are applied on the joint transmission. Since these friction mechanisms involves in lubricated film, it is appropriate to use Stribeck curve for quantification. This curve could describe the friction coefficient to be dependent on a parameter combining the velocity and the load. There are certain researches that proposed varied expression of the load-velocity friction model, in order to quantify robot manipulator joints.

In developing manipulation tasks, it is usually to assume a friction model that is appropriate for the particular task specified in order to make the problem tractable. The chosen model should be comprehensive enough to model the most relevant dynamics of the system, yet tangible enough so that it is feasible to formulate controllers, and moreover, the friction forces expected from the model are measurable by the robot's tactile and/or force sensors so as to verify and validate design.

Given a selected friction model, what remains unknown are the parameters that define the model and system. As discussed in preceding sections, there exist many uncertainties in friction models. There are several alternatives to deal with these parametric uncertainties of the friction model: assume a priori known friction parameters for the model; pre-manipulate the object prior to the task in order to identify the friction properties. This can be achieved by online estimating the friction parameters while performing the manipulation task. The research field of robot identification was explored decades ago [99–103]. In these works, use of the regressor form of the system equations of motion facilitated dynamic identification [100–103].

In the development of new robotic systems, robotic identification techniques have evolved to address the new challenges in modeling, analysis, system identification, and control design [103–147]. Many studies used a regressor form of a model of system. Some new methods offer more efficient identification than static measurements, in which it estimates parameters of actuator, drive train, and load efficiently. Some approaches represent the general Euler–Lagrange dynamics into the regressor form, and some approaches automatically derives the regressor form for given Euler–Lagrange equations of motion.

The dynamic model shown in Eqs. (4.23a) and (4.23b) has been used for parameter identification as well as simulation and control problems of the manipulators. In general, a given rigid body is described by multiple parameters such as the mass, the six independent entries of the inertia tensor, and the three coordinates of the center of mass. However, due to constraints and coupled kinematics, this number could be lower for a manipulator. An n-link robot then has a maximum of $10n$ dynamic parameters. Estimating the parameters from the design data of the manipulator is not simple, but there are a few techniques available.

In order to obtain accurate estimates of the dynamic parameters, it is possible to use an identification technique that exploits that the dynamic model are linear with respect to a suitable set of dynamic parameters. In Eqs. (4.23a) and (4.23b), term ψ_k is called the regressor and q is the parameter vector. Assuming that values of the joint positions q, velocities dq/dt, and accelerations d^2q/dt^2 can be recorded during execution of trajectories with the manipulator, and that the joint torques can be measured from sensors in the joints or current measurements, it is then possible to calculate the dynamic parameters directly by the parameterized system Eqs. (4.23) and (4.24). However, finding a minimal set of parameters that can parameterize the dynamic model is difficult in general, and as mentioned, measurements of q, dq/dt, d^2q/dt^2, during motion is a requirement for using this method.

Another approach is by simulation through computer-aided-design modeling. The various components of the manipulator are then modeled digitally on the basis of their geometry and type of materials, and features in the simulation system can be used to measure the parameters. Inaccuracies will occur with this technique, because of simplifications in the modeling and the loss of information about complex dynamic effects like joint friction. Dynamic simulation and model-based control of industrial robots require a correct description of the equations of motion and an accurate knowledge of the dynamic parameters. Determination of the dynamic parameters by using simulation data may not yield a complete representation as it is difficult to include dynamic parameters such as joint friction, masses introduced by additional equipment and stiffness parameters. Experimental parameter identification using the assembled robot may be the only practical method to determine accurate values of the dynamic parameters. The dynamic parameters can be identified by using a linear least squares method, provided that all degrees of freedom are measured during the identification experiment. Industrial robots are usually only equipped with motor encoders that measure joint angles. This implies that additional sensors are required to measure the joint degrees of freedom, e.g. by means of acceleration sensors, link position and/or velocity sensors or torque sensors. Some approach is restricted to linear models, implying that nonlinear terms such as Coulomb friction and gravitation forces cannot be included in the dynamic model. Omitting these forces is not realistic, since joint friction accounts for a significant part of the motor torque of industrial robots. Therefore, the application of additional sensors seems to be the most promising method to identify the dynamic parameters of manipulators, particularly for flexible-joint robots. Although joint friction is complicated in reality, a simple model, which is the combination of viscous and Coulomb friction, is normally used to describe the friction phenomenon for all joints.

Many systematic procedures for identification the dynamic model of robot manipulator have been proposed. There are certain main considerations needs to be addressed. For example, the validity of the most commonly used joint friction model with the combination of multiple factors such as viscous and Coulomb friction should be carefully considered. The problem of the physical feasibility of the identified parameters has to be addressed. In addition to the standard linear least-square estimator, varied constrained optimization problem have been used to obtain the identified parameters.

Friction plays a key role in robotic manipulation, and incorrect estimated friction coefficients can lead to malfunctions or system instability. In terms of academic research, a standard robot identification procedure consists of dynamic modeling, excitation trajectory design, data collection, signal preprocess, parameter identification, and model validation. Chae et al. [99] proposed the least-squares method to realize the estimation of dynamical parameters. Grotjahn et al. [112] used the two-step approach to perform the identification of robot dynamics. Gautier and Poignet [104] obtained a dynamical model of SCARA robot from experimental data with a weighted least-squares method. Behzad et al. [113] applied fractional subspace method to identify a robot model in simulation field. Wu et al. [115] proposes a closed-loop identification procedure using modified Fourier series as exciting trajectories. First, a static continuous friction model is involved to model joint friction for realizable friction compensation in controller design. Second, the method satisfying the boundary conditions are firstly designed as periodic exciting trajectories. To minimize the sensitivity to measurement noise, the coefficients of method are optimized according to the condition number criterion. Moreover, to obtain accurate parameter estimates, the maximum likelihood estimation method considering the influence of measurement noise is adopted. Figure 4.9 shows the flowchart of the model verification scheme.

Hamon et al. [119–121] proposed new inverse dynamic identification model for multiple degrees of freedom (DoF) serial robot, where the dry sliding friction force is a linear function of both the dynamic and the external forces, with a velocity-dependent coefficient. A new sequential identification procedure is carried out. At a first step, the friction model parameters are identified for each joint (1 DoF), moving one joint at a time. As a second step, these values are fixed in the n-DoF dynamic model for the identification of all robot inertial and gravity parameters. For the two steps, the identification concatenates all the joint data collected while the robot is tracking planned trajectories with different payloads to get a global

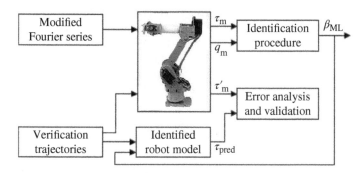

Figure 4.9 Model verification scheme [115].

least squares estimation of inertial and new friction parameters. An experimental validation is carried out with an industrial 3-DoF robot.

4.5.2 Identification of Parameters of Friction Models

This section presents a recently developed method [119–121]. For this case, all forces, positions, velocities, and accelerations have a conventional positive sign in the same direction. This defines a motor convention for the mechanical behavior. The dynamic model of a rigid robot manipulator Eqs. (4.23a) and (4.23b) composed of n moving links can be rewritten as follows:

$$M(q)\ddot{q} + C(q,\dot{q})\dot{q} + Q(q) = \tau + \tau_f + \tau_{off} + \tau_{ext} \tag{4.30}$$

$$\tau_{dyn} = \tau + \tau_f + \tau_{off} + \tau_{ext} \tag{4.31}$$

where $q, \dot{q}, \ddot{q}$ are, respectively, the $(n \times 1)$ vectors of generalized joint positions, velocities, and accelerations, $M(q)$ is the $(n \times n)$ robot inertia matrix, $C(q, \dot{q})$ is the $(n \times n)$ matrix of centrifugal and Coriolis effects, $Q(q)$ is the $(n \times 1)$ vector of gravitational forces. τ is the $(n \times 1)$ input torque vector on the motor side of the drive chain, without offset,

$$\tau = g_f v_f \tag{4.32}$$

where v_f is the $(n \times 1)$ vector of current references of the current amplifiers, g_f is the $(n \times n)$ matrix of the drive gains.

τ_f is the $(n \times 1)$ vector of the loss force due to viscous and dry frictions, without offset:

$$\tau_f = -F_v \dot{q} - F_C \, sign(\dot{q}) \tag{4.33}$$

where F_v is the $(n \times n)$ diagonal matrix of viscous parameters, F_C is the $(n \times n)$ diagonal matrix of dry friction parameters, and $sign(.)$ denotes the sign function. τ_{off} is an offset force that regroups the amplifier offset and the asymmetrical Coulomb friction coefficient. τ_{ext} is the $(n \times 1)$ external forces vector in the joint space.

Then Eq. (4.30) can be rewritten as the inverse dynamic model (IDM), which calculates the motor torque vector τ as a function of the generalized coordinates:

$$\tau = M(q)\ddot{q} + C(q,\dot{q})\dot{q} + Q(q) + F_C \, sign(\dot{q}) + F_V \dot{q} + \tau_{off} - \tau_{ext}$$
$$= \tau_{out} + F_C \, sign(\dot{q}) + F_V \dot{q} + \tau_{off} \tag{4.34}$$

where $\tau_{ext} = \tau_{dyn} - \tau_{out}$ is the output force (the load force) of the drive chain. The choice of the modified Denavit and Hartenberg frames attached to each link allows to obtain a dynamic model linear in relation to a set of standard dynamic parameters χ_{St}:

$$\tau = D_{St}(q, \dot{q}, \ddot{q})\chi_{St} \tag{4.35}$$

where $D_{St}(q, \dot{q}, \ddot{q})$ is the regressor and χ_{St} is the vector of the standard parameters that are the coefficients XXj, XYj, XZj, YYj, YZj, ZZj of the inertia tensor of link j denoted j, the mass of the link j called mj, the first moments vector of link j around the origin of frame j denoted $^jM_j = [MX_j MY_j MZ_j]^T$ the friction coefficients FVj, FCj, the actuator inertia called Iaj, and the offset $\tau offj$. The velocities and accelerations are calculated using well-tuned band pass filtering of the joint position.

The base parameters are the minimum number of parameters from which the dynamic model can be calculated. They are obtained by eliminating and by regrouping some standard inertial parameters. The minimal IDM can be written as

$$\tau = D(q, \dot{q}, \ddot{q})\chi \tag{4.36}$$

where $D(q, \dot{q}, \ddot{q})$ is the minimal regressor and χ is the vector of the base parameters. The IDM Eq. (4.35) is sampled while the robot is tracking a trajectory to get an overdetermined linear system such that:

$$Y(\tau) = W(q, \dot{q}, \ddot{q})\chi + \rho \tag{4.37}$$

with $Y(\tau)$ the measurements vector, W the observation matrix and ρ the vector of errors.

The LS solution $\hat{\chi}$ minimizes the two-norm of the vector of errors ρ. W is a $(r \times b)$ full rank and well-conditioned matrix where $r = N \times n$, with N the number of samples on the trajectories. The LS solution $\hat{\chi}$ is given by:

$$\hat{\chi} = (W^T W)^{-1} W^T Y = W^+ Y \tag{4.38}$$

It is calculated using the QR factorization of W. Standard deviations $\sigma_{\hat{\chi}_i}$ are estimated using classical and simple results from statistics. The matrix W is supposed to be deterministic, and ρ, a zero-mean additive independent noise, with a standard deviation such as:

$$C_{\rho\rho} = E(\rho\rho^T) = \sigma_\rho^2 I_r \tag{4.39}$$

where E is the expectation operator and I_ρ, the $(r \times r)$ identity matrix. An unbiased estimation of σ_ρ is:

$$\hat{\sigma}_\rho^2 = \|Y - W\hat{\chi}\|^2 / (r - b) \tag{4.40}$$

The covariance matrix of the standard deviation is calculated as follows:

$$C_{\hat{\chi}\hat{\chi}} = E[(\chi - \hat{\chi})(\chi - \hat{\chi})^T] = \sigma_\rho^2 (W^T W)^{-1} \tag{4.41}$$

$\sigma_{\hat{\chi}_i}^2 = C_{\hat{\chi}\hat{\chi}_{ii}}$ is the ith diagonal coefficient of $C_{\hat{\chi}\hat{\chi}}$. The relative standard deviation $\%\sigma_{\hat{\chi}_{ri}}$ is given by:

$$\%\sigma_{\hat{\chi}_{ri}} = 100\sigma_{\hat{\chi}_i}/\hat{\chi}_i \tag{4.42}$$

However, experimental data are corrupted by noise and error modeling and W is not deterministic. This problem can be solved by filtering the measurement vector Y and the columns of the observation matrix W.

In the following, the example in [145, 146] is presented, in which the following friction model is used

$$f_i(\dot{q}_i) = f_{vi}\dot{q}_i + sgn(\dot{q}_i)\left[f_{ki} + (f_{si} - f_{ki})e^{-\frac{|\dot{q}_i|}{k_i}}\right] \tag{4.43}$$

Since the friction model in Eq. (4.43) is nonlinear, it is very difficult to factorize it to be a liner parameter form. As such the parameter estimation is taken using following three steps. First the viscous friction and kinetic friction are estimated at high velocity. Then the static friction is estimated at steady-state using a step input. Finally, the empirical parameter ki of joint i is estimated at low velocity.

Step 1: *Estimate all linear parameters at high velocity.* At high velocity, Eq. (4.43) becomes

$$f_i \approx f_{vi}\dot{q} + sgn(\dot{q}_i)f_{ki} \tag{4.44}$$

Eq. (4.44) together with the elements in $M(q)$, $C(q, \dot{q})$, and $G(q)$ have linear-in-parameter property. They can be factorized into the form We, where. W is the regressor matrix, and e is the parameter vector to be estimated.

It is possible to apply standard recursive least square algorithm to estimate the parameters.

Step 2: *Estimate the static friction using step input.*

To estimate static friction, we move the robot arms to any valid position using step input. According to [145, 146], at steady state, we have

$$F_s + G = k_p(q_d - q) \tag{4.45}$$

where $F_s = [f_{s1}, f_{s2}, \cdots, f_{sn}]^T$ is the static friction vector. The vector G has been estimated in the previous step, then

$$\widehat{F}_s = k_p(q_d - q) - \widehat{G} \tag{4.46}$$

If all joints are exactly right up, the gravity torques are zero. Eq. (4.45) becomes

$$\widehat{F}_s = k_p \, (q_d - q) \tag{4.47}$$

Step 3: *Estimate the empirical constant k at low velocity,*

At low velocity, since all the parameters except ki are estimated and known, we have

$$y = sgn\,(\dot{q})\,(\widehat{f}_s - \widehat{f}_k)\ e^{\frac{-|\dot{q}|}{k}} \tag{4.48a}$$

$$k = \frac{-\,|\,\dot{q}\,|}{\ln \dfrac{y}{sgn(\dot{q})(\widehat{f}_s - \widehat{f}_k)}} \tag{4.48b}$$

$$0 < \frac{y}{sgn\,(\dot{q})\,(\widehat{f}_s - \widehat{f}_k)} < e \tag{4.48c}$$

$$y = \tau - \widehat{M}(q)\,\ddot{q} - \widehat{C}(q, \dot{q})\,\dot{q} - \widehat{f}_v \dot{q} - sgn\,(\dot{q})\,\widehat{f}_k - \widehat{G} \tag{4.48d}$$

where the hat (^) denotes the estimate value of the parameters, and fv, fk, G are the vector corresponding to the viscous friction, kinetic friction, the empirical parameter vector respectively, e is the base of the natural logarithm. Equation (4.48c) is necessary, because $k > 0$, From Eq. (4.48b), since the parameter k has a nonlinear contribution to the total friction force, it is difficult to estimate this parameter using traditional recursive least-squares algorithm. One solution is to obtain the value of k in each step during the estimation period at low velocity, then use the sum of the square of these values to calculate the quadratic mean.

From this friction torque estimation scheme, the friction parameters can be easily obtained by estimating the linear model first at high velocity. Then the static friction can be estimated using step input. Finally, the empirical parameter k is estimated at low velocity, where we can simplify the equation and apply the quadratic mean method.

Usually, the joint transmission friction model for robots is composed of a viscous friction force and of a constant dry sliding friction force. However, according to the Coulomb law, the dry friction force depends linearly on the load driven by the transmission, which has to be taken into account for robots working with large variation of the payload or inertial and gravity forces. Moreover, for robots actuating at low velocity, the Stribeck effect must be taken into account.

This chapter proposes a new inverse dynamic identification model for an n-DoF serial robot, where the dry sliding friction force is a linear function of both the dynamic and the external forces, with a velocity-dependent coefficient. A new sequential identification procedure is carried out. At a first step, the friction model parameters are identified for each joint (1 DoF), moving one joint at a time (this step has been validated in). At a second step, these values are fixed in the n-DoF dynamic model for the identification of all robot inertial and gravity parameters. For the two steps, the identification concatenates all the joint data collected while the robot is tracking planned trajectories with different payloads to get a global least squares estimation of inertial and new friction parameters. An experimental validation is carried out with an industrial 3-DoF robot.

For complex nonlinear model with multiple parameters, the intelligent algorithms such as GA and PSO have been used to identify the accurate parameters. Both GA and PSO are used to identify the static and dynamic parameters of the LuGre function models. In this kind of method, the displacements and controllable input force are obtained by sensors and the controllable input force is considered as the approximate value to the objective.

Some intelligence computation algorithms have been reported as a useful tool in robot model identification. The traditional genetic algorithm and the improved genetic algorithm have been used to identify robot model. However, while dealing with identification problems regarding complex and large-scale parameters, the GA algorithm would be stuck on local optimum.

Ding et al. [109] designed optimal periodic excitation trajectories to integrate the identification experiment, data collection, and signal preprocess. All the unknown parameters are well identified by artificial bee colony algorithm. When comparing the measured torques and the predicting torques, it was demonstrated that the proposed method can accurately estimate the robot dynamical parameters.

Further, a model validation has been carried out to verify the validity of the identified model.

4.5.3 Uncertainty Analysis

Friction uncertainties phenomena arise when two surfaces come into contact. They are a natural component of many manipulation tasks; for instance, friction forces may not only be present at the task level during manipulation, but also internally in the manipulator itself. The motors of the joints that compose robotic manipulators can exhibit significant static and dynamic friction effects, which need to be compensated appropriately in order to achieve accurate control of the manipulator. Even though friction has been extensively studied by the grasping, manipulation, and control community, it remains a formidable challenge in manipulation, mainly because it is difficult for the robot to know a priori (before manipulating the object) exactly what the friction properties of the object are. Paradoxically, these properties can actually only be measured once the robot actually manipulates the object.

The accurate and reliable quantification of friction coefficients remains challenging due to the dependence of friction coefficients on the material, surface, environment, and measuring equipment. Some research has critically examined the uncertainty associated with the determination of dynamic friction coefficient. In order to enable the confident use of friction, it is necessary to provide a quantitative, defensible statement regarding its reliability. The topic of data and model uncertainty has been addressed by researchers.

In [132, 142–144], researchers outlined how polynomial chaos theory (PCT) can be utilized for uncertain analysis of manipulator and controller design in a 4-DoF selective compliance assembly robot-arm-type manipulator with variation in both the link masses and payload. It includes a simple linear control algorithm into the formulation to show the capability of the PCT framework. Wiener's polynomial chaos (PC) presented a framework for separating stochastic elements of a system response from deterministic components, which is defined as a nonsampling-based technique to establish development of uncertainty in a dynamical system, when the system parameters have probabilistic uncertainty.

It can be employed for statistical analysis of dynamical systems since it allows probabilistic description of the uncertainty effects. Due to

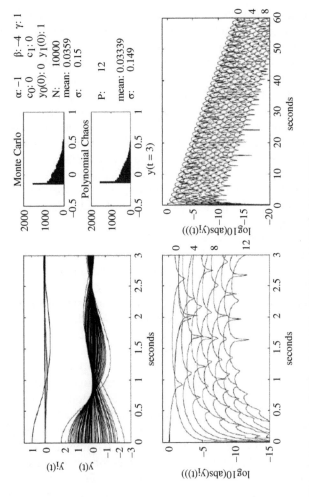

Figure 4.10 PC accurately generated the short-term probability density function (PDF).

low computational cost, it is an efficient alternative to Monte Carlo simulations. Figure 4.10 showed that (PCT) accurately generated the short-term probability density function and indicated the stability of the system response.

The design of a commercially produced mechanical system is influenced by the manufacturing variation that affects its performance. Suitable analysis tools are required to simulate and predict the potential dynamics generated by this variation. For conducting such uncertain dynamic and static analyses for a robot manipulator as shown in Figure 4.10, (PCT can be employed as a unifying framework. The trajectory control of redundant robotic arms is an important area of research that envisions efficient optimization algorithms. A robotic arm is termed as a kinematically "redundant manipulator" if it possesses higher DoF than required to establish any orientation and position of the end effector. In redundant manipulators, inverse kinematics present infinite solutions so they can be configured optimally for an assigned task.

Varghese et al. [76] demonstrated that a redundant robot controlled by a feedback linearization technique could display quasi-periodicity and chaos. The standard technique (e.g. closed-loop pseudo inverse control, Figure 4.10) suggested for solving their kinematics resulted in chaotic joint motions with erratic arm configurations. The dimension of their dynamic response was found to be fractal.

Quach and Liu [147] propose a Bayesian identification method where the robot manipulates the mechanism in different directions and classifies the kinematic model among a set of models such as revolute and sliding mechanisms and latches.

4.6   Friction Compensation and Control of Robot Manipulator Dynamics

The control problem of robot manipulator requires the definition of the input signals for the joints, such as torques or actuator input voltages, as so to realize a predefined behavior for the manipulator. The achievable performances for robot manipulator can be very different for many reasons such as the control techniques available to solve such a problem, the hardware used to implement the control algorithms, the real mechanical configuration of the robot, etc. The realized robot manipulator performance is mainly influenced by the mechanical system and actuation system design. Fundamentally, the problem defined by Eqs. (4.23a) and (4.23b) is a nonlinear stochastic dynamics problem.

For dynamics model Eq. (4.24), the control problem is as follows: Define the generalized force τ to be applied to the joint so as to obtain a desired trajectory $q_d(t)$. This is concerned with the friction compensation. Friction compensation is a very basic problem in motion control and therefore there exists many literatures on this topic [148–196]. The approaches include model-based friction compensation [153], sliding mode control [175, 176], classical integrator action/disturbance observers [154], and adaptive controllers [197–202], etc. Since the parameters of friction strongly vary with multiple factors such as position, velocity, load, and temperature as well as time, model-based friction compensation is quite inaccurate. Adaptive and sliding mode techniques adapt to parameter variations, are however sensitive to un-modeled dynamics. On the other hand, standard linear techniques such as integrators or disturbance observers are typically used in industrial robotic manipulator applications and show good practical performance. Their analysis, however, is usually based on linear techniques and does not really apply to the strongly nonlinear robotic systems. For integrators, only local convergence results have been achieved in robotic manipulators.

Compensating for friction has been implemented by many ways, for example, one approach is to use the variable structure systems approach: first, variable structure-based observers are developed for friction estimation in mechanical systems with or without information of velocity; the estimates are then used for a model-based feedforward compensation for friction. For a non-model-based approach, a robust sliding mode controller can also be used to cancel the influence of friction. Sigmoidal functions can be used to reduce chattering. Simulation results verify the validity of the proposed technique to compensate for friction of both static and dynamic models.

Bona and Indri [15] suggest analyzing the various friction compensation techniques by classifying them in terms of the adopted kind of control strategy:

1) A "fixed" friction compensation term is added to a more general control scheme, such as a joint independent one or an inverse dynamics control algorithm, by estimating the friction parameters off-line, on the basis of a chosen friction model, following proper

conventional identification procedures [153, 154, 158, 180]. If a dynamic friction model is chosen, an observer is then inserted to estimate the friction parameters. A further correction or a robust control action can be possibly added to the a priori estimated compensation term.

The non-model-based friction compensation examples include the classic PD and PID feedback controllers, impulsive control and dither. Although the position tracking with a PD control is stable, stick–slip even chaos could occur at low velocity. This effect could be eliminated using high derivative, high proportional feedback or a combination of the both. Another issue is the steady-state tracking error which can be solved with integral control. Integral control can cause limit cycling at low or zero velocity and can cause other difficulties at velocity reversals. These issues can be handled with a deadband and anti-windup at velocity reversal. However, these methods also introduce new issues like steady-state errors and decreasing performance. Impulsive control attains precise tracking by using series of small impacts. This makes combinations of other technologies possible. The use of the impulsive control can achieve a controlled breakaway and then another controller can take over to regulate the movements. Dither is a high frequency variation introduced into a system to modify dynamic behavior leading to a stabilizing effect. The focus in dithering is its capability to smoothen the discontinuity of friction at low velocity.

2) Only some main characteristics in the friction such as the stick peak are taken into account for compensation [159–161]. The identification of the required friction characteristics is determined off-line. However, the online compensation action is tuned based on the identification.

Some proposed system identification methods do not require special test benchmarks or the actuator being isolated. The parameters are identified using a trajectory relevant to the manipulator general operation. This ensures satisfied results in critical position, velocity, and acceleration ranges. Even if the Stribeck curve parameters still need to be tuned, all other parameters can be calculated using a single linear optimization.

3) Model-based friction compensation has been applied for friction compensation [5, 6, 162–167]. Model-based adaptive algorithms have been widely applied for on-line friction compensation. The adaptive schemes are based on a particular, static or dynamic,

friction model, whose parameters are tuned online to obtain a satisfying compensation action also when significant variations are present.

Model-based approaches to compensate friction can be subdivided into feedforward and feedback approaches. The feedback compensation uses the real-time sensed system output while the feedforward compensation operates with a desired reference.

Both approaches can be expanded with adaption mechanisms to cope with changing model parameters due to varying circumstances.

The feedback and feedforward friction compensations are briefed as follows. The most basic form of feedforward is shown in Figure 4.11. The feedback controller assures the stability and increases the disturbance rejection. The feedforward controller improves tracking performance.

Figure 4.11 shows the plant P, feedback controller C and feedforward controller F. The signals shown in Figure 4.11 are the reference trajectory r, servo error e, plant input u, plant output y, and feedforward signal f. The overall transfer function is given by,

$$To = yr = FP + CP = 1 + CP \tag{4.49}$$

To acquire a perfect tracking, a suitable F is needed such that $To = 1$. This yields

$$FP + CP = 1 + CP \tag{4.50}$$

Therefore,

$$F = 1/P \tag{4.51}$$

In some research, the feedforward controller consists of an acceleration feedforward part to compensate for the inertia forces and an IDM. In many practices, the uncertainties and un-modeled and non-minimum phase dynamics make this approach difficult to implement.

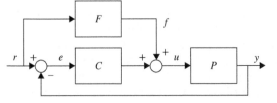

Figure 4.11
Schematic block diagram of elementary feedback and feedforward control.

A variation of this feedforward control is that the feedforward is given by a pre-filter R of the reference trajectory r,

$$R = 1 + FC \qquad (4.52)$$

This is used in combination with a non-causal filter. The effect of the prefilter R is same as the feedforward controller F.

Friction feedforward compensation pre-calculates the friction, in terms of the desired reference trajectory, which does not need any extra real-time closed-loop calculation.

In contrast to the feedback control, feedforward control will not affect the closed-loop stability. Before friction feedback compensation is applied, a closed-loop stability evaluation must be done so as to guarantee stability. The feedforward method is limited by the use of the reference.

The advantage of the feedback compensation is that it uses the measured state to calculate the friction compensation.

For example, a typical friction model used is the Karnopp model with viscous friction added. This static model only requires the velocity as input for both feedforward and feedback friction compensation. Some research applied the approach to allow the feedforward controller to provide accurate reference tracking and the feedback controller to reject disturbances. The term *disturbances* cover disturbances caused by external loads and un-modeled dynamics in this case.

First the model parameters are identified by using the feedback error signal. By adjusting the feedforward parameters, the uncompensated dynamics are extracted from the feedback errors. The friction model is used in both the feedforward and the feedback friction compensation. Both implementations can result in a similar improvement in terms of the position and velocity error compared to an uncompensated situation.

To design feedforward controllers for robots, a model that includes friction is critical, but friction model may have uncertainty. Existing research illustrated that it is possible to design feedforward controllers to allow the final position of the motion is robust to uncertainty in the friction model. It is possible to eliminate the sensitivity of the final state of motion to uncertainty in varied types of friction. Moreover, the feedforward controllers can be optimized for robustness to uncertainty in the friction model. The adaptive nonlinear friction compensation schemes were once proposed to handle different types of parametric uncertainty in the LuGre dynamic friction model, including non-uniform friction force variations and normal force variations. By using the estimated friction parameters and the dual observers for the unmeasurable friction state, adaptive nonlinear controllers can be designed to achieve globally asymptotic tracking of the given velocity reference signal.

1) Strategies that are not based on a particular friction model can be applied to counteract the friction effects, by properly choosing the control gain parameters or by using non-model-based observers [168–171].

2) Another group of friction compensation techniques takes advantage of the fuzzy, neural, and genetic algorithms to reconstruct the friction torques to be compensated or for a suitable self-tuning of the controller gains [176–178].

A few approaches for online learning of friction compensation torques exist. Some research apply reinforcement learning with a neural network to control a system with changing friction parameters. Iterative Learning Control has been proposed as a method for friction compensation before. It is also illustrated that a PD type iterative control can be used for learning torque commands to overcome friction effects on a fixed trajectory.

References

1 Karnopp, D. (1985). Computer simulation of stick-slip friction in mechanical systems. *ASME J. Dyn. Syst. Meas.Control* 107 (1): 100–103.

2 Dupont, P.E. (1990). Friction modeling in dynamic robot simulation. In: *Proceedings of 1990 IEEE International Conference on Robotics and Automation*, 1370–1376.

3 Armstrong-Helouvry, B. (1991). *Control of Machines with Friction*. Boston: MA, Kluwer Academic Publisher.

4 Armstrong-Helouvry, B., Dupont, P., and de Wit, C.C. (1994). A survey of models, analysis tools and compensation methods for the control of machines with friction. *Automatica* 30 (7): 1083–1138.

5 Canudas de Wit, C., Olsson, H., Astrom, K. et al. (1995). A new model for control of systems with friction. *IEEE Trans. Autom. Control* 40 (3): 419–425.

6 Olsson, H., Astrom, K., Canudas de Wit, C. et al. (1998). Friction models and friction compensation. *Eur. J. Control* 4: 176–195.

7 Hensen, R., Angelis, G., van de Molengraft, M. et al. (2000). Grey-box modelling of friction: an experimental case study. *Eur. J. Control* 6 (3): 258–267.

8 Swevers, J., Al-Bender, F., Ganseman, C. et al. (2000). An integrated friction model structure with improved pre-sliding behavior for accurate friction compensation. *IEEE Trans. Autom. Control* 45 (4): 675–686.

9 Lampaert, V., Swevers, J., and Al-Bender, F. (2002). Modification of the Leuven integrated friction model structure. *IEEE Trans. Autom. Control* 47 (4): 683–687.

10 Dupont, P., Hayward, V., Armstrong, B. et al. (2002). Single state elasto-plastic friction models. *IEEE Trans. Autom. Control* 47 (5): 787–792.

11 Lin, T.-Y., Pan, Y.-C., and Hsieh, C. (2003). Precision-limit positioning of direct drive systems with the existence of friction. *Control Eng. Pract.* 11: 233–244.

12 Ferretti, G., Magnani, G., and Rocco, P. (2004). Single and multistate integral friction models. *IEEE Trans. Autom. Control* 49 (12): 2292–2297.

13 Al-Bender, F., Lampaert, V., and Swevers, J. (2005). The generalized Maxwell-slip model: a novel model for friction simulation and compensation. *IEEE Trans. Autom. Control* 50 (11): 1883–1887.

14 Kermani, M., Wong, M., Patel, R. et al. (2004). Friction compensation in low and high-reversal-velocity manipulators. In: *Proc. ICRA'04, 2004 IEEE Int. Conf. on Robotics and Automation*, vol. 5, 4320–4325.

15 Bona, B. and Indri, M. (2005). Friction Compensation in Robotics: an Overview, 44th IEEE Conference on Decision and Control, and the European Control Conference, Seville, Spain, December 12–15.

16 Witono Susanto, F.L., Babuska, R., and van der Weiden, T. (2008). Adaptive friction compensation: application to a robotic manipulator, Proceedings of 17th World Congress in International Federation of Automatic Control.

17 Bittencourt, A.C., Wernholt, E., Sander-Tavallaey, S. et al. (2010). An extended friction model to capture load and temperature effects in robot joints, 2010 IEEE/RSJ International Conference on Intelligent Robots and Systems (IROS), 18–22 October, Taiwan.

18 Bittencourt, A.C. and Gunnarsson, S. (2012). Static friction in a robot joint Modeling and identification of load and temperature effects. *ASME J. Dyn. Syst. Meas. Control* 134 (5).

19 Bittencourt, A.C., Saarinen, K., Sander, S. et al. (2011). *A Data-Driven Method for Monitoring of Repetitive Systems: Applications to Robust Wear Monitoring of a Robot Joint.* Technical report from Automatic Control at Linköpings universitet.

20 Caccavale, F., Cilibrizzi, P., Pierri, F. et al. (2009). Actuators fault diagnosis for robot manipulators with uncertain model. *Control Eng. Pract.* 17 (1): 146–157.

21 Namvar, M. and Aghili, F. (2009). Failure detection and isolation in robotic manipulators using joint torque sensors. *Robotica* 28 (4): 549–561.

22 McIntyre, M., Dixon, W., Dawson, D. et al. (2005). Fault identification for robot manipulators. *IEEE Trans. Rob.* 21 (5): 1028–1034.

23 Vemuri, A.T. and Polycarpou, M.M. (2004). A methodology for fault diagnosis in robotic systems using neural networks. *Robotica* 22 (04): 419–438.

24 Brambilla, D., Capisani, L., Ferrara, A. et al. (2008). Fault detection for robot manipulators via second-order sliding modes. *IEEE Trans. Ind. Electron.* 55 (11): 3954–3963.

25 Strom, K.J.A. and Canudas-de Wit, C. (2008). Revisiting the LuGre friction model. *IEEE Control Syst. Mag.* 28 (6): 101–114.

26 Avraham Harnoy, B.F.S.C. (2008). Modeling and measuring friction effects. *IEEE Control Syst. Mag.* 28 (6).

27 Al-Bender, F. and Swevers, J. (2008). Characterization of friction force dynamics. *IEEE Control Syst. Mag.* 28 (6): 64–81.

28 Mattone, R. and Luca, A.D. (2009). Relaxed fault detection and isolation: an application to a nonlinear case study. *Automatica* 42 (1): 109–116.

29 Drinčić, B., Tan, X., and Bernstein, D.S. (2011). Why are some hysteresis loops shaped like a butterfly? *Automatica* 47: 2658–2664.

30 Drinčić, B., and Bernstein, D.S. (2011). A sudden-release bristle model that exhibits hysteresis and stick-slip friction. Proc. Amer. Contr. Conf., San Francisco, CA: 2456–2461.

31 Lock, J. and Dupont, P.E. (2011). Friction modeling in concentric Tube Robots. *IEEE Int Conf on Robot Autom.* 1139–1146.

32 Majdou, F., Perret-Liaudet, J., Belin, M. et al. (2015). Decaying law for the free oscillating response with a pseudo-polynomial friction law: analysis of a super low lubricated friction test. *J. Sound Vib.* 348: 263–281.

33 Vigué, P., Vergez, C., Karkar, S. et al. (2017). Regularized friction and continuation: comparison with Coulomb's law. *J. Sound Vib.* 389: 350–363.

34 Graf, M. and Ostermeyer, G.-P. (2015). Friction-induced vibration and dynamic friction laws: instability at positive friction–velocity-characteristic. *Tribol. Int.* 92: 255–258.

35 Hu, Y., Wang, L., Politis, D.J. et al. (2017). Development of an interactive friction model for the prediction of lubricant break-down behavior during sliding wear. *Tribol. Int.* 110: 370–377.

36 Ruderman, M. (2017). On break-away forces in actuated motion systems with nonlinear friction. *Mechatronics* 44: 1–5.

37 Khan, Z.A., Chako, V., and Nazir, H. (2017). A review of friction models in interacting joints for durability design. *Friction* 5 (1): 1–22.

38 Waiboer, R., Aarts, R., and Jonker, B. (2005). Velocity dependence of joint friction in robotic manipulator with gear transmission. In: *Multibody Dynamics 2005, ECCOMAS Thematic Conference, Madrid* (ed. J.M. Goicolea, J. Cuadrado and J.C. García Orden), 21–24.

39 Waiboer, R. (2007). *Dynamical Modeling, Identification, and Simulation of Industrial Robots.* the Netherlands: Netherlands Institute for Metals Research https://research.utwente.nl/en/publications/dynamic-modelling-identification-and-simulation-of-industrial-rob.

40 Xu, H., Kahraman, A., Anderson, N.E. et al. (2007). Prediction of mechanical efficiency of parallel-Axis gear pairs. *ASME J. Mech. Design* 129 (1): 58–68.

41 Xu, H. (2005). Development of a Generalized Mechanical Efficiency Prediction Methodology. PhD dissertation, The Ohio State University.

42 Simoni, L., Beschi, M., Legnani, G. et al. (2015). Friction Modeling with Temperature Effects for Industrial Robot Manipulators, 2015 IEEE/RSJ International Conference on Intelligent Robots and Systems (IROS), Sept. 28– October 2. Hamburg, Germany.

43 Liu, Y.F., Li, J., Zhang, Z.M. et al. (2015). Experimental comparison of five friction models on the same test-bed of the micro stick-slip motion system. *Mech. Sci.* 6: 15–28.

44 Tijani, B. and Akmeliawati, R. (2012). Support vector regression based friction modeling and compensation in motion control system. *J. Eng. Appl. Artif. Intell. Arch.* 25 (5): 1043–1052.

45 Skowronski, J. (2012). *Control Dynamics of Robotic Manipulators.* Elsevier Science. ISBN: 0323158110.

46 Jazar, R.N. (2010). *Theory of Applied Robotics, Kinematics, Dynamics, and Control*, 2e. New York: Springer. ISBN: 978-1-4419-1749-2.

47 Lewis, F.L., Dawson, D.M., and Abdallah, C.T. (2004). *Robot Manipulator Control Theory and Practice*, 2e. Marcel Dekker, Inc. ISBN: 0-8247-4072-6.

48 Peshkin, M.A. (1990). *Robotic Manipulation Strategies.* Prentice Hall. ISBN: 0-13-781493-3.

49 Flores, P. and Lankarani, H.M. (2016). *Contact Force Models for Multibody Dynamics.* Springer. ISBN: 978-3-319-30896-8.

50 Niku, S.B. (2011). *Introduction to Robotics: Analysis, Control, Applications*, 2e. Wiley. ISBN: 978-0470604465.

51 Melchiorri, C. (2008). *Dynamic Model of Robot Manipulators.* Universit'a di Bologna http://www-lar.deis.unibo.it/people/cmelchiorri/Files_Robotica/FIR_05_Dynamics.pdf.

52 Sciavicco, L. and Siciliano, B. (2000). *Modelling and Control of Robot Manipulators*, 2e. Springer. ISBN: 1852332212.

53 Goicolea, J.M., Cuadrado, J., Orden J.C. García (eds.), (2005). Multibody Dynamics, ECCOMAS Thematic Conference Madrid, Spain, 21–24 June.

54 Murray, R.M., Li, Z.S., and Shankar, S. (1994). *A Mathematical Introduction to Robotic Manipulation.* CRC Press.

55 Bajd, T., Mihelj, M., Lenar, J. et al. (2010). *Robotics.* Springer Verlag.

56 Hollerbach, J.M. (2007). A recursive lagrangian formulation of manipulator dynamics and a comparative study of dynamics formulation complexity. *IEEE Trans. Syst. Man Cybern.* 10 (11): 730–736.

57 Sciavicco, L. and Siciliano, B. (2000). *Modelling and Control of Robot Manipulators.* Springer Verlag.

58 Spong, M.W., Hutchinson, S., and Vidyasagar, M. (2006). *Robot Modeling and Control.* Hoboken, NJ: Wiley.

59 Siciliano, B., Sciavicco, L., Villani, L. et al. (2009). Robotics. Modelling, planning and control. In: *Advanced Textbooks in Control and Signal Processing*. Springer-Verlag.

60 Spong, M.W., Lewis, F.L., and Abdallah, C.T. (1993). Robot Control. In: *Dynamics, Motion Planning, and Analysis*. New York: IEEE Press.

61 Lee, C.S.G., González, R.C., and Fu, K.S. (2006). *Tutorial on Robotics*, 2e. Silver Springs.

62 Nehmzow, U. *Scientific Methods in Mobile Robotics: Quantitative Analysis of Agent Behavior*. Springer.

63 Nehmzow, U. (2009). *Robot Behaviour: Design, Description, Analysis and Modelling*. Springer.

64 Schoner, G., Dose, M., and Engels, C. (1995). Dynamics of behavior: theory and applications for autonomous robot architectures. *Rob. Auton. Syst.* 16: 213–245.

65 Chung, W., Li-Chen, F., and Hsu, S.-H. (2008). Motion control. In: *Springer Handbook of Robotics* (ed. B. Siciliano and O. Khatib). ISBN: 978-3-540-23957-4.

66 Moberg, S. (2010). Modeling and Control of Flexible Manipulators, Linköping studies in science and technology. Dissertations. 1349.

67 Spong, M.W. (1989). On the force control problem for flexible joint manipulators. *IEEE Trans. Autom. Control* 34: 107–111.

68 Jung, J., Penning, R., Ferrier, N.J. et al. (2011). A Modeling Approach for Continuum Robotic Manipulators: Effects of Nonlinear Internal Device Friction, IEEE/RSJ International Conference on Intelligent Robots and Systems, September 25–30. San Francisco.

69 Liu, X., Li, H., Wang, J. et al. (2006). Dynamics analysis of flexible space robot with joint friction. *IEEE Trans. Control Syst. Technol.* 14 (2).

70 Korayem, M.H. and Rahimi, H.N. (2011). Nonlinear dynamic analysis for elastic robotic arms, 6. *Front. Mech. Eng.* (2): 219–228.

71 de Lima, J.J., Tusset, A.M., Janzen, F.C. et al. (2016). SDRE applied to position and vibration control of a robot manipulator with a flexible link. *J. Theor. Appl. Mech.* 54 (4): 1067–1078.

72 Schwarz, M. and Behnke, S. (2013). Compliant Robot Behavior using Servo Actuator Models identified by Iterative Learning

Control, Proceeding of 17th RoboCup International Symposium, Eindhoven, Netherlands.

73 Zulfatman, M.M. and Mardiyah, N.A. (2017). Two-Link Flexible Manipulator Control Using Sliding Mode Control Based Linear Matrix Inequality, IAES International Conference on Electrical Engineering, Computer Science and Informatics, IOP Conf. Series: Materials Science and Engineering 1149506 (72809107) 012008 Engineering 190.

74 Fonseca Ferreira, N.M., Duarte, F.B., Lima, M.F.M. et al. (2008). Application of fractional Calculus in the dynamical analysis and control of mechanical manipulators. *Fract. Calc. Appl. Anal.* 11 (1): 91–112.

75 Mahout, V., Lopez, P., Carcasses, J.P. et al. (1993). Complex behaviors of a two revolute joints robot: harmonic, subharmonic, higher harmonic, fractional harmonic, chaotic responses. Proceedings of International Conference on Systems, Man and Cybernetics, Systems Engineering in the Service of Humans.

76 Varghese, M., Fuchs, A., and Mukundan, R. (1991). Chaotic zero dynamics in kinematically redundant robots. *IEEE Trans. Aerosp. Electron. Syst.* 27: 784–796.

77 Ravishankar, A.S. and Ghosal, A. (1996). Possible chaotic motions in a feedback controlled 2R robot. In: *Proceedings of the 13th IEEE International Conference on Robotics and Automation,* 1241–1246.

78 Ravishankar, A.S. and Ghosal, A. (1997). Chaos in robot control equations. *Int. J. Burfurcat. Chaos* 7 (3): 707–720.

79 Ravishankar, A.S. and Ghosal, A. (1999). Nonlinear dynamics and chaotic motions in feedback-controlled two- and three-degree-of-freedom robots. *Int. J. Rob. Res.* 18 (1): 93–108.

80 Verduzco, F. and Alvarez, J. (1999). Bifurcation analysis of a 2-DoF robot manipulator driven by constant torques. *Int. J. Bifurcat. Chaos Appl. Sci. Eng.* 9 (4): 617–627.

81 Jackson, E.A. and Grosu, I. (1995). An open-plus-closed-loop (OPCL) control of complex dynamic systems. *Physica D* 85 (1–2): 1–9.

82 Chen, L.-Q. and Liu, Y.-Z. (1998). A modified open-plus-closed-loop approach to control chaos in nonlinear oscillations. *Phys. Lett. A* 245 (1–2): 87–90.

83 Chen, L.-Q. (2001). An open-plus-closed-loop control for discrete chaos and hyperchaos. *Phys. Lett. A* 281 (5–6): 327–333.

84 Chen, L.-Q. and Liu, Y.-Z. (2002). An open-plus-closed-loop approach to synchronization of chaotic and hyperchaotic maps. *Int. J. Bifurcat. Chaos* 12 (5): 1219–1225.

85 Chen, L.-Q. (2004). The parametric open-plus-closed-loop control of chaotic maps and its robustness. *Chaos, Solitons and Fractals* 21 (1): 113–118.

86 Tian, Y.-C., Tadé, M.O., and Tang, J. (2000). Nonlinear open-plus-closed-loop (NOPCL) control of dynamic systems. *Chaos, Solitons and Fractals* 11 (7): 1029–1035.

87 Verduzco, F. and Alvarez, J. (1999). Bifurcation analysis of a 2-DoF robot manipulator driven by constant torques. *Int. J. Bifurcat. Chaos Appl. Sci. Eng.* 9 (4): 617–627.

88 Duarte, F.B.M. and Machado, J.A.T. (2000). Chaos dynamics in the trajectory control of redundant manipulators. Proceedings, ICRA'00. IEEE International Conference on Robotics and Automation.

89 Duarte, F.B.M. and Machado, J.A.T. (2000). Motion chaos in the pseudoinverse control of redundant robots. Proceedings, 6th International Workshop on Advanced Motion Control.

90 Verduzco, F. and Alvarez, J. (2000). Homoclinic Chaos in 2-DoF robot manipulators driven by PD controllers. *Nonlinear Dyn.* 21 (2): 157–171.

91 Paar, V., Pavin, N., Paar, N. et al. (2000). Nonlinear dynamics of a single-degree robot model, part 2: onset of chaotic transients. *Robotica* 18: 201–208.

92 da Graca Marcos, M., Duarte, F., and Tenreiro Machado, J.A. (2008). Fractional dynamics in the trajectory control of redundant manipulators. *Commun. Nonlinear Sci. Numer. Simul.* 13: 1836–1844.

93 Gribovskaya, E., Khansari-Zadeh, S.M., and Billard, A. (2010). Learning nonlinear multivariate dynamics of motion in robotic manipulators. *Int. J. Rob. Res.* 30 (1): 80–117.

94 David, S.A., Balthazar, J.M., Julio, B.H.S. et al. (2012). The fractional-nonlinear robotic manipulator: AIP Conference Proceedings on Modeling and dynamic simulations (1): 298.

95 Han, Q.K., Zhao, X.-Y., and Wen, B.-C. (2008). Synchronization motions of a two-link mechanism with an improved OPCL method. *Appl. Math. Mech.* 29 (12): 1561–1568.

96 Han, Q.K., Qin, Z., Yang, X. et al. (2007). Rhythmic swing motions of a two-link robot with a neural controller. *Int. J. Innov. Comput. Inform. Control* 3 (2): 335–342.

97 Han, Q.K., Zhao, X., Yang, X. et al. (2010). Periodic and chaotic motions of a two-bar linkage with OPCL controller. *Math. Prob. Eng.* 2010: 1–15.

98 Zang, X., Iqbal, S., Zhu, Y. et al. (2016). Applications of chaotic dynamics in robotics. *Int. J. Adv. Rob. Syst.* 13: 60–71.

99 An, C.H., Atkeson, C.G., and Hollerbach, J.M. (1985). Estimation of inertial parameters of rigid body links of manipulators. 24th IEEE Conference on Decision and Control, 24: 990–995.

100 Gautier, M. and Khalil, W. (1988). On the identification of the inertial parameters of robots. Proceedings of the 27th IEEE Conference on Decision and Control, 3: 2264–2269.

101 Gautier, M. and Khalil, W. (1989). Identification of the minimum inertial parameters of robots. Proceedings of 1989 IEEE International Conference on Robotics and Automation 3: 1529–1534.

102 Gautier, M. and Khalil, W. (1990). Direct calculation of minimum set of inertial parameters of serial robots. *IEEE Trans. Rob. Autom.* 6 (3): 368–373.

103 Gautier, M. and Khalil, W. (1991). Exciting trajectories for the identification of base inertial parameters of robots. Proceedings of the 30th IEEE Conference on Decision and Control, 1: 494–499.

104 Gautier, M. and Poignet, P. (2001). Extended Kalman filtering and weighted least squares dynamic identification of robot. *Control Eng. Pract.* 9 (12): 1361–1372.

105 Calaore, G., Indri, M., and Bona, B. (2001). Robot dynamic calibration: optimal excitation trajectories and experimental parameter estimation. *J. Rob. Syst.* 18 (2): 55–68.

106 Olsen, M.M. and Petersen, H.G. (2001). A new method for estimating parameters of a dynamic robot model. *IEEE Trans. Rob. Autom.* 17 (1): 95–100.

107 van Zutven, P. (2009). Modeling, identification and stability of a humanoid robot. MS thesis, Eindhoven University of Technology.

108 Vuong, N.D. and Ang, M.H. Jr. (2009). Dynamic model identification for industrial robots. *Acta Polytech. Hung.* 6 (5): 51–68.

109 Li, D., Hongtao, W., Yao, Y. et al. (2015). Dynamic model identification for 6-DoF industrial robots. *J. Rob.* 2015: 1–9.

110 Wu, J., Wang, J., and You, Z. (2010). An overview of dynamic parameter identification of robots. *Rob. Comput. Integ. Manufact.* 26 (5): 414–419.

111 Atkeson, C.G., An, C.H., and Hollerbach, J.M. (1986). Estimation of inertial parameters of manipulator loads and links. *Int. J. Rob. Res.* 5 (3): 101–119.

112 Grotjahn, M., Daemi, M., and Heimann, B. (2001). Friction and rigid body identification of robot dynamics. *Int. J. Solids Struct.* 38 (10): 1889–1902.

113 Behzad, H., Shandiz, H., Noori, T.A. et al. (2011). Robot identification using fractional subspace method. In: *Proceedings of the 2nd International Conference on Control, Instrumentation and Automation (ICCIA'11)*, 1193–1199. Iran: Shiraz.

114 Kostic, D., Steinbuch, M., Hensen, R. et al. (2004). Modeling and identification for high-performance robot control: an RRR-robotic arm case study. *IEEE Trans. Control Syst. Technol.* 12 (6).

115 Wu, W., Zhu, S., Wang, X. et al. (2012). Closed-loop dynamic parameter identification of robot manipulators using modified Fourier series. *Int. J. Adv. Rob. Syst.* doi: 10.5772/45818.

116 Liu, Z., Huang, P., and Zhenyu, L. (2016). Recursive differential evolution algorithm for inertia parameter identification of space manipulator. *Int. J. Adv. Rob. Syst.* 13: 104.

117 Kinsheel, A., Taha, Z., Deboucha, A. et al. (2012). Robust least square estimation of the CRS A465 robot arm's dynamic model parameters. *J. Mech. Eng. Res.* 4 (3): 89–99.

118 Yan, D., Lu, Y., and Levy, D. (2015). Parameter identification of robot manipulators: a heuristic particle swarm search approach. *PLoS One* 10 (6): e0129157.

119 Hamon, P., Gautier, M. and Garrec, P. (2011). New Dry Friction Model with Load- and Velocity-Dependence and Dynamic Identification of Multi-DoF Robots, 2011 IEEE International Conference on Robotics and Automation. May 9–13, Shanghai, China.

120 Hamon, P., Gautier, M. and Garrec, P. (2010). Dynamic Identification of Robot with a Dry Friction Model Depending on Load and Velocity. IEEE Int. Conf. on Intelligent Robots and Systems, Taipei, Taiwan.

121 Hamon, P., Gautier, M. Garrec, P. (2010). Dynamic Modeling and Identification of Joint Drive with Load-Dependent Friction Model. IEEE Int. Conf. on Advanced Intelligent Mechatronics, Montreal, Canada.

122 Bona, B., Indri, M. and Smaldone, N. (2003). *Nonlinear friction estimation for Digital Control of Direct-Drive Manipulators*.

123 van Zutven, P., Kosti, D. and Nijmeijer H. (2010). Parameter identification of robotic systems with series elastic actuators, 8th IFAC Symposium on Nonlinear Control Systems University of Bologna, Italy, September 1–3.

124 Schwarz, M. and Behnke, S. (2013). Compliant Robot Behavior using Servo Actuator Models identified by Iterative Learning Control, Proceeding of 17th RoboCup International Symposium, Eindhoven, Netherlands.

125 Hardeman, T., Aarts, R. and Jonker, B. (2005). Modeling and identification of robots with both joint and drive flexibilities, 5th International Conference on Multibody Systems, Nonlinear Dynamics, and Control, Long Beach, CA, USA, September 24–28.

126 Waiboer, R., Aarts, R., and Jonker, B. (2005). Modelling and Identification of a Six Axes Industrial Robot. ASME 2005 International Design Engineering Technical Conferences and Computers and Information in Engineering Conference, 6: 2265–2274.

127 Bittencourt, A.C. and Axelsson, P. (2014). Modeling and experiment Design for Identification of wear in a robot joint under load and temperature uncertainties based on friction data. *IEEE/ASME Trans. Mechatron* 19 (5): 1694–1706.

128 Hamon, P., Gautier, M., and Garrec, P. (2011). Dynamic Identification of Robots with a Dry Friction Model Depending on Load and Velocity, HAL Id: hal-00583177 https://hal.archives-ouvertes .fr/hal-00583177.

129 Zheng, Y.Q. (2012). Parameter identification of LuGre friction model for robot joints. *Advanced Materials Research Online* 479–481: 1084–1090.

130 Kammerer, N. and Garrec, P. (2013). Dry friction modeling in dynamic identification for robot manipulators: Theory and experiments, 2013 IEEE International Conference on Mechatronics, February, Vicenza, Italy, 422–429.

131 Gomes, S.C.P. and Santos da Rosa, V. (2003). A new approach to compensate friction in robotic actuators, Proceedings of ICRA '03. IEEE International Conference on Robotics and Automation.

132 Indri, M., Trapani, S., and Lazzero I. (2016). Development of a general friction identification framework for industrial manipulators, IECON 2016 42nd Annual Conference of Industrial Electronics Society, IEEE.

133 Gogoussis A. and Donath, M. (1988). Coulomb friction effects on the dynamics of bearings and transmissions in precision robot mechanisms. Proceedings of IEEE International Conference on Robotics and Automation, Philadelphia, Pennsylvania, 3: 1440–1446.

134 Zheng, Y.Q. (2012). Parameter identification of LuGre friction model for robot joints. *Adv. Mater. Res.* 479–481: 1084–1090.

135 Indri, M., Lazzero, I. Antoniazza, A. et al. (2013). Friction modeling and identification for industrial manipulators, IEEE 18th Conference on Emerging Technologies & Factory Automation (ETFA).

136 Omata, T. and Nagata, K. (2000). Rigid body analysis of the indeterminate grasp force in power grasps. *IEEE Trans. Rob. Autom.* 16 (1): 46–54.

137 Cai, L., and Song, G. (1993). A smooth robust nonlinear controller for robot manipulators with joint stick-slip friction, Proceedings of 1993 IEEE International Conference on Robotics and Automation.

138 Morel, G., Iagnemma, K., and Dubowsky, S. (2000). The precise control of manipulators with high joint-friction using base force/torque sensing. *Automatica* 36: 931–941.

139 Tafazoli, S., de Silva, C.W., and Lawrence, P.D. (1995). Friction estimation in a planar electrohydraulic manipulator. Proceedings of the 1995 American Control Conference.

140 Liu, G., Iagnemma, K., Dubowsky, S. et al. (1998). A Base Force/Torque Sensor Approach to Robot Manipulator Inertial Parameter Estimation, Proceedings of the 1998 IEEE International Conference on Robotlcs & Automation Leuven, Belgium.

141 Denkena, B., Heimann, B., Abdellatif, H. et al. (2005). Design, modeling and advanced control of the innovative parallel manipulator PaLiDA, Proceedings, 2005 IEEE/ASME International Conference on Advanced Intelligent Mechatronics.

142 Suzuki, T., Miyoshi, W., and Nakamura, Y. (2013). Friction modeling and identification for industrial manipulators, 2013 IEEE 18th Conference on Dynamic Performance, Emerging Technologies & Factory Automation (ETFA), 10–13 September.

143 Indri, M., Lazzero, I., Antoniazza, A. et al. (2009). Friction modeling and identification for industrial manipulators of a SCARA robot manipulator with uncertainty using polynomial Chaos theory. *IEEE Trans. Rob.* 25 (1): 21–30.

144 Hover, F.S. and Triantafyllou, M.S. (2006). Application of polynomial chaos in stability and control. *Automatica* 42: 789–795.

145 Voglewede, P.A., and Monti, A. (2006). Variation Analysis of a Two Link Planar Manipulator Using Polynomial Chaos Theory. ASME 2006 International Design Engineering Technical Conferences and Computers and Information in Engineering Conference. American Society of Mechanical Engineers.

146 Voglewede, P., Smith, A.H.C., and Monti, A. (2009). Dynamic performance of a SCARA robot manipulator with uncertainty using polynomial chaos theory. *IEEE Trans. Rob.* 25: 206–210.

147 Quach, N.H. and Liu, M. (2004). *Friction Torques Estimation and Compensation for Robot Arms*. NH Quach http://semanticscholar .org.

148 Quach, N.H. and Liu, M. (2000). A 3-Step Set-Point Control Algorithm for Robot Arms. Proc. of IEEE Int. Conf. on Robotics and Automation, 1296–1301.

149 Barragán, P.R., Kaelbling, L.P. and Lozano-Pérez, T. (2013). Interactive Bayesian Identification of Kinematic Mechanisms, http://lis .csail.mit.edu/pubs/barragan-icra14.pdf

150 Le Tien, L. Albu-Schäffer, A., De Luca, A. et al. (2008). Friction Observer and Compensation for Control of Robots with Joint Torque Measurement, 2008 IEEE/RSJ International Conference on Intelligent Robots and Systems Acropolis Convention Center Nice, France.

151 Yesildirak, A., Lewis, F.L., and Jagannatha, S. (1999). *Neural Network Control of Robot Manipulators and Nonlinear Systems*. Taylor & Francis.

152 Llama, M.A., Kelly, R., and Santibanez, V. (2000). Stable computed torque control of robot manipulators via fuzzy self-tuning. *IEEE Trans. Syst. Man Cybern. Part B* 30 (1): 143–150.

153 Velez-Diaz, D. and Tang, Y. (2004). Adaptive robust fuzzy control of nonlinear systems. *IEEE Trans. Syst. Man Cybern.Part B* 34 (3): 1596–1601.

154 Popovic, M.R., Gorinevsky, D.M., and Goldenberg, A.A. (2000). High-precision positioning of a mechanism with nonlinear friction using a fuzzy logic pulse controller. *IEEE Trans. Control Syst. Technol.* 8 (1): 151–158.

155 Papadopoulos, E. and Chasparis, G. (2002). Analysis and model based control of servomechanisms with friction. In Proc. IEEE/RSJ Int. Conf. on Intelligent Robots and System, 3: 2109–2114.

156 Putra, D., Moreau, L., and Nijmeijer, H. (2004). Observer-based compensation of discontinuous friction. In Proc. 43rd IEEE Conf. on Decision and Control: 4940–4945.

157 Moreno, J., Kelly, R., and Campa, R. (2003). Manipulator velocity control using friction compensation. *IEE Proceed. Control Theory Appl.* 150 (2): 119–126.

158 Moreno, J., and Kelly, R. (2003). Manipulator velocity field control with dynamic friction compensation. In Proc. 42nd IEEE Conf. on Decision and Control 4: 3834–3839.

159 Slotine, J.-J. and Li, W. (1988). Adaptive manipulator control: A case study. *IEEE Trans. Autom. Control* 33 (11): 995–1003.

160 Shiriaev, A. Robertsson, and R. Johansson, (2003). Friction compensation for passive systems based on the LuGre model. In Proc. 2nd IFAC Workshop on Lagrangian and Hamiltonian Methods for Nonlinear Control, Seville, Spain: 183–188.

161 Southward, S., Radcliffe, C., and MacCluer, C. (1991). Robust nonlinear stick-slip friction compensation. *ASME J. Dyn. Syst. Meas. Control* 113 (4): 639–645.

162 Kang, M.S. (1998). Robust digital friction compensation. *Control Eng. Pract.* 6: 359–367.

163 Mei, Z.-Q. Xue, Y.-C. Zhang, G.-L. et al. (2003). The nonlinear friction compensation in the trajectory tracking of robot. In Proc. 2003 Int. Conf. on Machine Learning and Cybernetics, 4: 2457–2462.

164 Vedagarbha, P., Dawson, D., and Feemster, M. (1999). Tracking control of mechanical systems in the presence of nonlinear dynamic friction effects. *IEEE Trans. Control Syst. Technol.* 7 (4): 446–456.

165 Zhu, Y. and Pagilla, P. (2002). Static and dynamic friction compensation in trajectory tracking control of robots. In Proc. ICRA'02 IEEE Int. Conf. on Robotics and Automation): 2644–2649.

166 Feemster, M., Vedagarbha, P., Dawson, D. et al. (1999). Adaptive control techniques for friction compensation. *Mechatronics* 9: 125–145.

167 Hung, N., Tuan, H., Narikiyo, T. et al. (2002). Adaptive controls for nonlinearly parameterized uncertainties in robot manipulators. In Proc. 41st IEEE Conf. on Decision and Control, 2: 1727–1732.

168 Liu, G. (2002). Decomposition-based friction compensation of mechanical systems. *Mechatronics* 12: 755–769.

169 Liu, G., Goldenberg, A., and Zhang, Y. (2004). Precise slow motion control of a direct-drive robot arm with velocity estimation and friction compensation. *Mechatronics* 14: 821–834.

170 Tataryn, P., Sepehri, N., and Strong, D. (1996). Experimental comparison of some compensation techniques for the control of manipulators with stick-slip friction. *Control Eng. Pract.* 4 (9): 1209–1219.

171 Armstrong, B., Neevel, D., and Kusik, T. (2001). New results in NPID control: tracking, integral control, friction compensation and experimental results. *IEEE Trans. Control Syst. Technol.* 9 (2): 399–406.

172 Chen, W.-H., Ballance, D.J., Gawthrop, P.J. et al. (2000). A nonlinear disturbance observer for robotic manipulators. *IEEE Trans. Ind. Electron.* 47 (4): 932–938.

173 Ryu, J.-H., Song, J., and Kwon, D.-S. (2001). A nonlinear friction compensation method using adaptive control and its practical application to an in-parallel actuated 6-DoF manipulator. *Control Eng. Pract.* 9: 159–167.

174 Craig, J.J. (1988). *Adaptive Control of Mechanical Manipulators*. New York: Addison-Wesley Inc.

175 Xia, Q.H., Lim, S.Y., Ang, M.H. Jr., and Lim, T.M. (2004). Adaptive Joint Friction Compensation Using a Model-Based Operational Space Velocity Observer. In: *IEEE Int. Conf. on Robotics and Automation*, 3081–3086.

176 Tomei, P. (2000). Robust adaptive friction compensation for tracking control of robot manipulators. *IEEE Trans. Autom. Control* 45: 2164–2169.

177 Ryu, J.-H., Song, J., and Kwon, D.-S. (2001). A nonlinear friction compensation method using adaptive control and its practical application to an in-parallel actuated 6-DoF manipulator. *Control Eng. Pract.* 9: 159–167.

178 Llama, M.A., Kelly, R., and Santibañez, V. (2000). Stable computedtorque control of robot manipulators via fuzzy self-tuning. *IEEE Trans. Sys. Man Cybern. Part B* 30 (1): 143–150.

179 Vélez-Díaz, D. and Tang, Y. (2004). Adaptive robust fuzzy control of nonlinear systems. *IEEE Trans. Syst. Man Cybern. Part B* 34 (3): 1596–1601.

180 Popovíc, M.R., Gorinevsky, D.M., and Goldenberg, A.A. (2000). High-precision positioning of a mechanism with nonlinear friction using a fuzzy logic pulse controller. *IEEE Trans. Control Syst. Technol.* 8 (1): 151–158.

181 Parra-Vega, V. and Arimoto, S. (1996). A passivity based adaptive sliding mode position-force control for robot manipulators. *Int. J. Adapt. Control Signal Process.* 10 (4–5): 365–377.

182 Chung, S.-K., Imayoshi, H., Hanamoto, T. et al. (2000). Implementation of frictionless robot manipulator using observer based sliding mode control. In: *Proc. 26th Annual Conf. of the IEEE Industrial Electronics Society*, vol. 1, 578–583. IECON.

183 de Wit, C.C., Olsson, H., Astron, K.J. et al. (1994). A new model for control of systems with friction IEEE Trans. *Autom. Control* 40: 419–425.

184 Kaloust, J.H. and Qu, Z. (1993). Robust guaranteed cost control of uncertain nonlinear robotic system using mixed minimum time and quadratic performance index. In: *Proc. 32nd IEEE Conf. on Decision and Control*, 1634–1635.

185 Lin, S. and Wang, S.-G. (2000). Robust control with pole clustering for uncertain robotic systems. *Int. J. Control Intel. Syst.* 28 (2): 72–79.

186 Wang, S.-G., Lin, S.B., Shieh, L.S. et al. (1998). Observer-based controller for robust pole clustering in a vertical strip and disturbance rejection in structured uncertain systems. *Int. J. Rob. Nonlinear Control* 8 (3): 1073–1084.

187 Jalaik, V. and Sharma, B.B. (2015). Observer based nonlinear control of robotic manipulator using Backstepping approach. *Rob. Auton. Syst.* 70: 83–91.

188 Plooij, M., Wolfslag, W., and Wisse, M. (2015). Robust feedforward control of robotic arms with friction model uncertainty. *Acta Astronaut.* 111: 1–18.

189 Liu, X.-F., Li, H.-Q., Chen, Y.-J. et al. (2015). Dynamics and control of space robotic considering joint friction. *Acta Astronaut.* 111 (1–18): 47.

190 van Geffen, V. (2015). *Friction Compensation in a Controlled One-Link Robot Using a Reduced-Order Observer*. Technische Universiteit Eindhoven Department Mechanical Engineering Dynamics and Control Technology Group.

191 Han, Q., Zhao, X., Li, X. et al. (2011). Bifurcations of a controlled two bar linkage motion with considering viscous frictions. *Shock Vib.* 18: 365–375.

192 Matsuoka, K., Ohyama, N., Watanabe, A. et al. (2005). Control of a giant swing robot using a neural oscillator. *Lect. Notes Comput. Sci.* 3611: 274–284.

193 Lu, R. (2006). *Synchronized Trajectory Tracking Control for Parallel Manipulators*. Doctorial Thesis of University of Toronto.

194 Nijmeijer, H. and Mareels, I.M.Y. (1997). An observer looks at synchronization. *IEEE Trans. Circ. Syst. Fundam. Theory Appl.* 44: 882–890.

195 Johnson, C.T. and Lorenz, R.D. (1991). Experimental identification of friction and its compensation in precise, position controlled mechanisms, Conference Record of the 1991 IEEE Industry Applications Society Annual Meeting.

196 Olsson, H. and Astrom, K.J. (1996). Observer-based friction compensation, Proceedings of the 35th IEEE Conference on Decision and Control.

197 Vedagarbha, P., Dawson, D.M., and Feemster, M. (1997). Tracking control of mechanical systems in the presence of nonlinear dynamic friction effects, Proceedings of the 1997 American Control Conference

198 Ha, Q.P., Bonchis, A., Rye, D.C. et al. (2000). Variable structure systems approach to friction estimation and compensation. Proceedings. ICRA 2000. IEEE International Conference on Robotics and Automation, 24–28 April, San Francisco, CA.

199 Zhang, G. and Furusho, J. (1997). Control of robot arms using joint torque sensors. Proceedings 1997 IEEE International Conference on Robotics and Automation, 25–25 April, Albuquerque, NM.

200 Brogliato, B. (1989). Adaptive Friction Compensation in Robot Manipulators: Low Velocities, Proceedings of 1989 IEEE International Conference on Robotics and Automation.

201 Tan, Y. and Kanellakopoulos, I. (1999). Adaptive nonlinear friction compensation with parametric uncertainties. Proceedings of the 1999 American Control Conference.

202 Nganga-Kouya, D., Saad, M., and Okou, F.A. (2011). A novel adaptive hybrid force-position control of a robotic manipulator. *Int. J. Model. Identif. Control* 13 (1–2): 1–17.

5

Force Feedback and Haptic Rendering

5.1 Overview of Robot Force Feedback

Robot force feedback generally provides the user in a teleoperation system the force feeling scaled with that of the slave robot via the haptic device. Also, the force feeling may come from the virtual robot in a virtual environment. Haptic devices are useful for tasks where visual information is not sufficient and may induce unacceptable manipulation errors, such as surgery or teleoperation in radioactive/chemical environments. The aim of haptic devices is to provide the user with a realistic feeling of the situation [1].

Haptic devices can be divided into many types [2], such as hand-controllers with force feedback, force-feedback gloves, exoskeleton haptic devices, and vibrotactile devices. In this chapter, we mainly discuss hand-controllers with force feedback.

Hand-controllers with force feedback can be divided into serial, parallel, and composed types. The Phantom product series from 3D Systems in the USA are a common serial force feedback device (see Figure 5.1). Parallel force feedback services are represented by products from Force Dimension in Switzerland, which are developed on a parallel delta mechanism (see Figure 5.2). The number of degree of freedoms (DoF) of motion or force present at the device body interface (the number of dimensions characterizing the possible movements or forces exchanged between device and operator) depends on the application.

A hand-controller can be viewed as having two basic functions: (i) To measure the positions and contact forces of the user's hand (and/or other body parts); (ii) To display contact forces and positions to the user. In some instances, contact forces and positions rendered to the

Dynamics and Control of Robotic Manipulators with Contact and Friction, First Edition.
Shiping Liu and Gang (Sheng) Chen.

Figure 5.1 Phantom omni haptic device.

Figure 5.2 Delta6 haptic device.

user need to be based on the position of the user's hand (and/or other body parts). For example, in VR, the user is manipulating his avatar. The mechanisms to measure the positions and contact forces are not discussed in this chapter. We only discuss how to present contact forces and tactile feeling to the user. The tactile feeling is also related to the friction between the user's hand and the surface of the object in addition to the pressure distribution.

The components of the serial force feedback service are not spatially overlapped, crossed, or parallel to each other, facilitating incorporation of other components and design. However, there are many singularity points in the workspace and the transmission performance of the mechanism near the singular point is poor, and the driving components may be damaged. The numerous cantilever structures of the series mechanism make the rigidity of the mechanism poor.

The parallel-type force feedback device has a high rigidity, strong bearing capacity, small moment of inertia and high precision, but its motion space is small. In a complex motion mechanism, the friction of the motion pair increases obviously and is coupled, which is very difficult to control.

It is difficult to control parallel mechanisms in real time, as required with haptics, because the forward kinematics is highly nonlinear and is often not in closed form. Moreover, the gravitational counterbalancing required for the hand-controller to appear transparent to the user, leads to software-operated static balancing, which increases the complexity and the time consumption of the control scheme. To overcome these difficulties, fast hardware capabilities are required [1].

Some hybrid force feedback services are developed to combine the advantages of serial and parallel types. Sometimes, a special task space is also required for the hybrid type [2].

If the slave robot works in the real world, the force acting on the end of the slave robot can be measured by force sensors installed on the slave robot. Some force sensors can measure up to six dimensional forces (e.g., three transitional forces and three torques). If the hand-contoller is used to control the robot in a virtual environment, the force information should be calculated in real time. That real-time calculation is called a haptic rendering. In haptic rendering, most algorithms are limited to point-based interaction between the user and the objects in the environment.

5.2 Generating Methods of Feedback Force

Depending on the number of DoF of the force feedback device, the direction of the required force is different. For a single DoF, only the force in the specified direction is generated. For a 2-DoF device, the force along the plane or tangent to the curve is generated. For a 3-DoF device, the three-dimensional force needs to be generated. For higher DoF, torques along one or more axes are often added on the basis of three-dimensional forces. Generally, one actuator is required for each DoF. The force acting on the operator is jointly generated by torques on each input motor.

5.2.1 Serial Mechanism

After the space force (torque) of the user is calculated, or after sensed, the torque of the motor acting on each DoF must be calculated according to the mechanics principles of the force feedback device. Generally, the manipulating end of the force feedback service moves slowly. To simplify the calculation, we can use static mechanics.

The mechanical structure of the serial force feedback device uses different mechanisms. The calculations of forward and inverse kinematics, and the calculations of forward and inverse dynamics, are naturally not the same. Here, we take the Phantom Premium 1.5 haptic device from 3D Systems (see Figure 5.3) as an example for analysis. The schematic diagram of the mechanism of the Phantom Premium 1.5 is shown in Figure 5.4. The architecture can be visualized as a 3-bar Revolute-Revolute-Revolute (RRR) spatial manipulator [3, 4].

5.2.1.1 Kinematics
The Denavit-Hartenberg (D-H) parameters are shown in Table 5.1 [5]. The transformation matrices of these three links are expressed as follows.

The homogeneous transformation matrix of the end-effector frame with respect to the fixed frame is then given by:

$$T = T_3 T_2 T_1$$

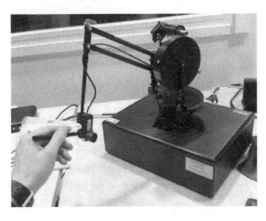

Figure 5.3 Phantom Premium 1.5 from 3D systems.

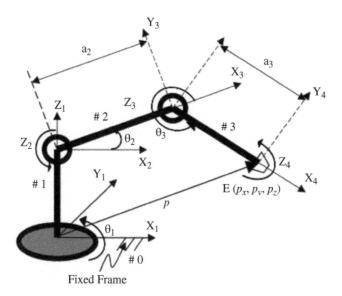

Figure 5.4 Schematic diagram of mechanism of the Phantom Premium 1.5.

Table 5.1 D-H parameters of Phantom Premium 1.5.

Link #	b_i	$\theta_i(JV)$	a_i	α_i
1	0	θ_1	0	$\pi/2$
2	0	θ_2	a_2	0
3	0	θ_3	a_3	0

b_i is the joint offset; θ_i is the joint angle; a_i is the link length; and α_i is the twist angle.

The coordinates of $P(P_x, P_y, P_z)$ are calculated as:

$$\begin{bmatrix} P_x \\ P_y \\ P_z \\ 1 \end{bmatrix} = T \begin{bmatrix} 0 \\ 0 \\ 0 \\ 1 \end{bmatrix} = \begin{bmatrix} a_2 \cos\theta_2 \cos\theta_1 + a_3 \cos\theta_1 \cos\theta_3 \\ a_2 \cos\theta_2 \sin\theta_1 + a_3 \sin\theta_1 \cos\theta_3 \\ a_1 + a_2 \sin\theta_2 - a_3 \cos\theta_3 \\ 1 \end{bmatrix} \qquad (5.1)$$

5.2.1.2 Inverse Kinematics

Since this is a 3-DoF manipulator, the inverse kinematics problem is to find the set of $(\theta_1, \theta_2, \theta_3)$ triples that move the manipulator to the desired end-effector position $P(P_x, P_y, P_z)$.

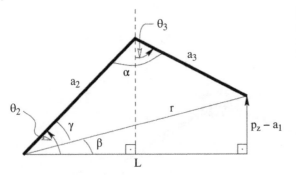

Figure 5.5 The side view of schematic mechanism.

θ_1 can be calculated as:

$$\theta_1 = ctg\left(\frac{P_y}{P_x}\right)$$

If $P_x == 0$, then $\theta_1 = \frac{\pi}{2}$.

Next, we calculate θ_2. In Figure 5.5, $\theta_2 = \gamma + \beta$. γ can be calculated by the law of cosine. In the upper triangle,

$$a_2^2 + r^2 + 2ra_2\cos\gamma = a_3^2$$

$$\gamma = \sqrt{L^2 + (P_z - a_1)^2},$$

$$l = \sqrt{P_x^2 + P_y^2}$$

Then,

$$\gamma = arccos\left(\frac{a_3^2 - a_2^2 - r^2}{2ra_2}\right)$$

The lower triangle is a right angle, so,

$$\beta = ctg\left(\frac{P_z - a_1}{\sqrt{P_x^2 + P_y^2}}\right)$$

So,

$$\theta_2 = \gamma + \beta = arccos\left(\frac{a_3^2 - a_2^2 - r^2}{2ra_2}\right) + ctg\left(\frac{P_z - a_1}{\sqrt{P_x^2 + P_y^2}}\right) \qquad (5.2)$$

In Figure 5.5, it can be seen that:

$$l = a_2 \cos\theta_2 + a_3 \sin\theta_3$$

So,

$$\theta_3 = \arcsin\left(\frac{l - a_2\cos\theta_2}{a_3}\right) = \arcsin\left(\frac{\sqrt{P_x^2 + P_y^2} - a_2\cos\theta_2}{a_3}\right)$$

(5.3)

5.2.1.3 Dynamics

The forward kinematic model of velocity is just the equation below:

$$\begin{bmatrix} \dot{x} \\ \dot{y} \\ \dot{z} \end{bmatrix} = J \begin{bmatrix} \dot{\theta}_1 \\ \dot{\theta}_2 \\ \dot{\theta}_3 \end{bmatrix}$$

where J is Jacobi matrix $J = \begin{bmatrix} J_{11} & J_{12} & J_{13} \\ J_{21} & J_{22} & J_{23} \\ J_{31} & J_{32} & J_{33} \end{bmatrix}$

$$\begin{bmatrix} \dot{x} \\ \dot{y} \\ \dot{z} \end{bmatrix} = \begin{bmatrix} \dfrac{\delta x}{\delta t} \\ \dfrac{\delta y}{\delta t} \\ \dfrac{\delta z}{\delta t} \end{bmatrix}, \begin{bmatrix} \dot{\theta}_1 \\ \dot{\theta}_2 \\ \dot{\theta}_3 \end{bmatrix} = \begin{bmatrix} \dfrac{\delta\theta_1}{\delta t} \\ \dfrac{\delta\theta_2}{\delta t} \\ \dfrac{\delta\theta_3}{\delta t} \end{bmatrix}$$

$$\therefore \begin{bmatrix} \delta x \\ \delta y \\ \delta z \end{bmatrix} = J \begin{bmatrix} \delta\theta_1 \\ \delta\theta_2 \\ \delta\theta_3 \end{bmatrix}$$

Let F be the force vector acting on the handle gripped by the operator, and τ be the torque vector applied to these three joint actuators, then the equation according to virtual work is as follows [5, 6]:

$$F^T \begin{bmatrix} \delta x \\ \delta y \\ \delta z \end{bmatrix} = \tau^T \begin{bmatrix} \delta\theta_1 \\ \delta\theta_2 \\ \delta\theta_3 \end{bmatrix}$$

$$F^T J \begin{bmatrix} \delta\theta_1 \\ \delta\theta_2 \\ \delta\theta_3 \end{bmatrix} = \tau^T \begin{bmatrix} \delta\theta_1 \\ \delta\theta_2 \\ \delta\theta_3 \end{bmatrix}$$

$$F^T J = \tau^T$$

$$\therefore \tau = J^T F$$

Generally, $F = \begin{bmatrix} f_x \\ f_y \\ f_z \end{bmatrix}$ is the virtual force on the contact point.

$$\therefore \tau = \begin{bmatrix} \tau_1 \\ \tau_2 \\ \tau_3 \end{bmatrix} = \begin{bmatrix} J_{11} & J_{21} & J_{31} \\ J_{12} & J_{22} & J_{32} \\ J_{13} & J_{23} & J_{33} \end{bmatrix} \begin{bmatrix} F_x \\ F_y \\ F_z \end{bmatrix}$$

$$\tau = \begin{bmatrix} \tau_1 \\ \tau_2 \\ \tau_3 \end{bmatrix} = \begin{bmatrix} J_{11}F_x + J_{21}F_y + J_{31}F_z \\ J_{12}F_x + J_{22}F_y + J_{32}F_z \\ J_{13}F_x + J_{23}F_y + J_{33}F_z \end{bmatrix} \tag{5.4}$$

5.2.2 Parallel Mechanism

There are many different types of parallel haptic devices, and many have more than 3 DoF. Here, we use a manipulandum device developed by the Huazhong University of Science and Technology as an example. It uses a delta parallel mechanism. It achieves three dimensional translational motions and generates feedback force on the hand of the operator. Its three-dimensional model is shown in Figure 5.6.

5.2.2.1 Kinematics Model

The kinematics model is shown in Figure 5.7. Input parameters are rotation angles of each motor, $\theta_i (i = 1, 2, 3)$. The output parameters are the coordinates of the origin P of the moving table frame, (P_x, P_y, P_z). Kinematic analysis of this delta mechanism–based hand-controller is to research the relationship between the input and output parameters of the mechanism and the calculation method. In the kinematic analysis, the parallelogram mechanism is simplified as a rod, and rods are simplified as the mass concentrating in the mass center.

In Figure 5.7 $A_i B_i C_i$ ($i = 1, 2, 3$) are kinematic chains. This delta mechanical mechanism consists of these three identical kinematic chains. $\Delta A_1 A_2 A_3$ and $\Delta C_1 C_2 C_3$ are regular triangles. Point O is the origin of the static platform frame, $O - XYZ$. $P - xyz$ is the frame of the upper moving platform. The Y-axis is perpendicular to the line $A_1 A_2$. The X axis is parallel to the line $A_1 A_2$. The Y axis is

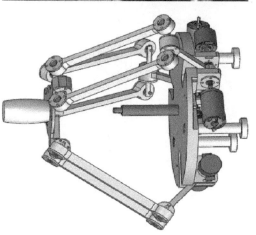

Figure 5.6 The 3D model of a delta manipulandum haptic device.

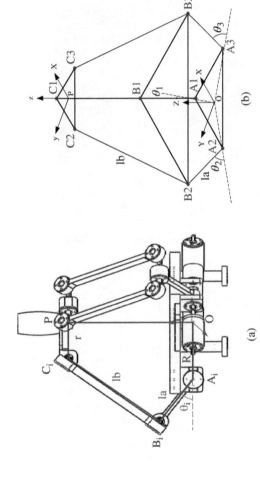

Figure 5.7 Kinematics model of this manipulandum: (a) definitions of parameters; (b) simplified kinematics model.

Table 5.2 Kinematic parameters of the delta mechanism based hand-controller.

Input parameters	$(\theta_1, \theta_2, \theta_3)$	Output parameters	(P_x, P_y, P_z)
Fixed frame	O-XYZ	Moving frame	P-xyz
Rod lengths	Rod A_2:l_a	Equivalent rod length of the parallelogram mechanism: l_b	
Radii of platforms	Static platform: R	Moving platform: r	Radii difference: Δr

perpendicular to the line $C_1 C_2$. The x axis is parallel to the line $C_1 C_2$. R is the upper static platform radius. r is the lower moving platform radius, $\Delta r = R - r$. l_a is the effective length of rod A_2, and l_b is the equivalent rod length of the parallelogram mechanism. Kinematic parameters of this hand-controller are listed in Table 5.2.

5.2.2.2 Forward Kinematics

The forward kinematics problem can be described as solving the output parameters, (P_x, P_y, P_z) given the input parameters $(\theta_1, \theta_2, \theta_3)$. The encoder measures the motor rotation angle, $(\theta_1, \theta_2, \theta_3)$, and calculates the coordinate of the origin of the moving frame, (P_x, P_y, P_z) through the forward kinematic algorithm, and obtains the displacement of the moving platform and then output the displacement value to control a manipulator.

The forward kinematic model is shown in Figure 5.8. In this model, three lines are drawn parallel to $B_i C_i$ ($i = 1, 2, 3$), respectively, all passing through the origin P of the moving frame and with the same length as with $B_i C_i$. The other end points are denoted as D_1, D_2, and D_3, respectively. The points P, D_1, D_2, and D_3 constitute tetrahedron shown in Figure 5.9.

Figure 5.10 shows the general position of this delta mechanism-based hand-controller. $\overrightarrow{PC_i} = \overrightarrow{D_i B_i}, \overrightarrow{B_i C_i} = \overrightarrow{PD_i} (i = 1, 2, 3)$. In the tetrahedron, D_1, D_2, and D_3 point coordinates can be solved according to the vector relationship. All edge lengths of this tetrahedron are known.

So, the forward kinematics problem is transformed into solving the vertex coordinates given the vertices coordinates of the tetrahedron

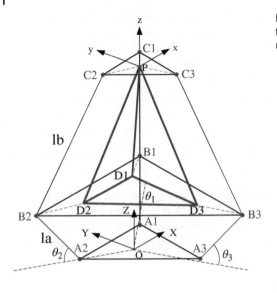

Figure 5.8 The forward kinematic model.

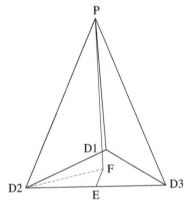

Figure 5.9 The tetrahedron of $PD_1D_2D_3$.

and the lengths of all edges. The following equation can be derived from vector relationships:

$$\overrightarrow{OD_i} = \overrightarrow{OB_i} - \overrightarrow{D_iB_i} = \overrightarrow{OB_i} - \overrightarrow{PC_i}$$
$$= \begin{pmatrix} (R - r + l_a \cos\theta_i)\cos\eta_i \\ (R - r + l_a \cos\theta_i)\sin\eta_i \\ l_a \sin\theta_i \end{pmatrix} \qquad (i = 1, 2, 3)$$

Figure 5.10 The general position of the manipulandum.

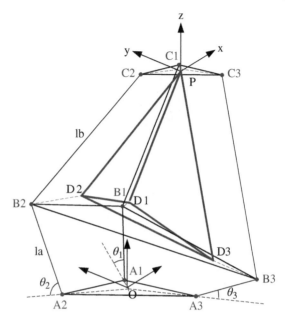

where η_i is defined as the angle between OA_i and positive direction of the X-axis.

$$\eta_i = \frac{4i - 3}{6}\pi, (i = 1, 2, 3)$$

In Figure 5.9, let F be the circumcenter of triangle $\Delta D_1 D_2 D_3$. Draw a perpendicular line to $D_2 D_3$, with the foot denoted as E. Since $\|\overrightarrow{PD_1}\| = \|\overrightarrow{PD_2}\| = \|\overrightarrow{PD_3}\|, \overrightarrow{D_2 D_3} \perp \Delta PEF$, and $\overrightarrow{D_2 D_3} \perp \overrightarrow{FP}$. Likewise, $\overrightarrow{D_1 D_2} \perp \overrightarrow{FP}$ can be also derived. So, $\overrightarrow{FP} \perp \Delta D_1 D_2 D_3$ is obtained. Likewise, $\overrightarrow{OF} \perp \Delta D_1 D_2 D_3$ can also be derived. The directions of vectors $\overrightarrow{FP}$ and $\overrightarrow{OF}$ are the same as the normal direction of $\Delta D_1 D_2 D_3$. Let the unit normal vector of $\Delta D_1 D_2 D_3$ be denoted as $\vec{n}$.

Then, $\vec{n} = \dfrac{\overrightarrow{D_1 D_2} \times \overrightarrow{D_1 D_3}}{\|\overrightarrow{D_1 D_2}\| \cdot \|\overrightarrow{D_1 D_3}\|}$. So $\overrightarrow{FP} = \|\overrightarrow{FP}\| \cdot \vec{n}$ and $\overrightarrow{OP} = \|\overrightarrow{OP}\| \cdot \vec{n}$.

The following equations can be derived:

$$\|\overrightarrow{FP}\| = \sqrt{\|OD_2\|^2 - \|D_1 F\|^2}$$

$$\|\overrightarrow{OP}\| = \sqrt{\|OD_2\|^2 - \|D_1 F\|^2}$$

So, $\overrightarrow{FP} = \|\overrightarrow{FP}\| \cdot \overrightarrow{n} = \sqrt{\|OD_2\|^2 - \|D_iF\|^2} \cdot \overrightarrow{n} = \dfrac{\overrightarrow{D_1D_2} \times \overrightarrow{D_1D_3}}{\|\overrightarrow{D_1D_2}\| \cdot \|\overrightarrow{D_1D_3}\|}$

$$\overrightarrow{OP} = \|\overrightarrow{OP}\| \cdot \overrightarrow{n} = \|\overrightarrow{OP}\| = \sqrt{\|OD_2\|^2 - \|D_iF\|^2} \cdot \overrightarrow{n}$$

$$= \dfrac{\overrightarrow{D_1D_2} \times \overrightarrow{D_1D_3}}{\|\overrightarrow{D_1D_2}\| \cdot \|\overrightarrow{D_1D_3}\|} \tag{5.5}$$

Based on $\overrightarrow{OP} = \overrightarrow{OF} + \overrightarrow{FP}$, the coordinates (P_x, P_y, P_z) can be solved.

5.2.2.3 Inverse Kinematics

The inverse kinematics problem can be described as solving the input parameters, $(\theta_1, \theta_2, \theta_3)$ given the output parameters, (P_x, P_y, P_z).

In the fixed frame $O - XYZ$, the coordinates of point A_i can be described by vector $\overrightarrow{OA_i}$.

$$\overrightarrow{OA_i} = \begin{pmatrix} R \cos \eta_i \\ R \sin \eta_i \\ 0 \end{pmatrix},$$

The coordinates of point C_i in the frame of the moving platform are described by $\overrightarrow{PC_i}$.

$$\overrightarrow{PC_i} = \begin{pmatrix} r \cos \eta_i \\ r \sin \eta_i \\ 0 \end{pmatrix}$$

The coordinates of point B_i in the fixed frame $O - XYZ$ is described by the vector $\overrightarrow{OB_i}$.

$$\overrightarrow{OB_i} = \begin{pmatrix} (R + l_a \cos \theta_i) \cos \eta_i \\ (R + l_a \cos \theta_i) \sin \eta_i \\ l_a \sin \theta_i \end{pmatrix}$$

(P_x, P_y, P_z) are the coordinates of P in the fixed frame $O - XYZ$, so $\overrightarrow{OP} = (P_x, P_y, P_z)^T$. Vector $\overrightarrow{OC_i}$ in $O - XYZ$ is described as $\overrightarrow{OC_i} = \begin{pmatrix} r \cos \eta_i + P_x \\ r \sin \eta_i + P_y \\ P_z \end{pmatrix}$. $\overrightarrow{B_iC_i} = \overrightarrow{OC_i} - \overrightarrow{OB_i}$, so

$$\overrightarrow{B_iC_i} = \begin{pmatrix} r \cos \eta_i + P_x - (R + l_a \cos \theta_i) \cos \eta_i \\ r \sin \eta_i + P_y - (R + l_a \cos \theta_i) \sin \eta_i \\ P_z - l_a \sin \theta_i \end{pmatrix}$$

According to $|B_i C_i| = l_b$, Eq. (5.6) can be derived:

$$P_x^2 + P_y^2 + P_z^2 - 2P_x \cos\eta_i(\Delta r + l_a \cos\theta_i)$$
$$-2P_y \sin\eta_i(\Delta r + l_a \cos\theta_i)$$
$$-2P_z l_a \sin\theta_i + \Delta r^2 + 2\Delta r l_a \cos\theta_i + l_a^2 - l_b^2 = 0 \qquad (5.6)$$

It can be transformed into:

$$2P_z l_a \sin\theta_i + 2l_a \cos\theta_i(P_x \cos\eta_i + P_y \sin\eta_i - \Delta r) + l_b^2 - l_a^2$$
$$-P_x^2 - P_y^2 - P_z^2 - (\Delta r)^2 + 2\Delta r(P_x \cos\eta_i + P_y \sin\eta_i) = 0 \qquad (5.7)$$

Let $m_i = 2P_z l_a$, $n_i = 2l_a(P_x \cos\eta_i + P_y \sin\eta_i - \Delta r)$, and

$$p_i = l_b^2 - l_a^2 - P_x^2 - P_y^2 - P_z^2 - (\Delta r)^2 + 2\Delta r(P_x \cos\eta_i + P_y \sin\eta_i),$$

So Eq. (5.7) can be simplified as

$$m_i \sin\theta_i + n_i \cos\theta_i + p_i = 0 \qquad (5.8)$$

Squaring Eq. (5.8) and simplifying it allows Eq. (5.9) to be derived:

$$(m_i^2 - p_i^2)\tan^2\theta_i + 2m_i n_i \tan\theta_i + n_i^2 - p_i^2 = 0 \qquad (5.9)$$

Let $A_i = m_i^2 - p_i^2$, $B_i = 2m_i n_i$, $C_i = n_i^2 - p_i^2$, and $t_i = \tan\theta_i$, so Eq. (5.10) is obtained:

$$A_i t_i^2 + B_i t_i + C_i = 0 \qquad (5.10)$$

The solution for this equation is:

$$t_i = \frac{-B_i \pm \sqrt{B_i^2 - 4A_i C_i}}{2A_i} \qquad (5.11)$$

So,

$$\theta_i = \arctan t_i \qquad (5.12)$$

So, $(\theta_1, \theta_2, \theta_3)$ can be solved according to the coordinates of P in the fixed frame $O - XYZ$, (P_x, P_y, P_z). Sometimes, there may be several solutions, which can be selected in these two ways:

1) Subtract the corresponding value of previous moment from the current angle value, and take a group with a smaller difference.
2) Substitute the current angle value into the forward kinematics algorithm, solve the coordinates of the origin of the moving frame, compare the difference, then choose the one which yields the given value.

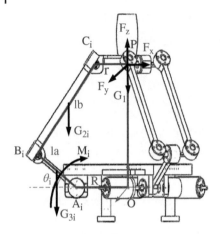

Figure 5.11 The static mechanics model of the whole hand-controller.

5.2.2.4 Dynamics Based on Virtual Work

A haptic input device can be used with relatively slow human hand motion during teleoperation of a remote manipulator or simulation of a procedure in virtual reality. During the movement of the hand-controller, the load mainly includes the operator's operating force (sometimes, that is also the same feedback force that the hand-controller exerts on the operator's hand conversely), the motor output torque, and the mechanical components of gravity. In this section, Coriolis forces, centrifugal forces, and frictions are temporally neglected.

Figure 5.11 shows the static model of this hand-controller. This hand-controller is in static equilibrium under the application of these forces.

The definition of each point and frame in the static mechanics model of the hand-controller are the same as the kinematic model. The moving gravity includes the moving platform gravity G_1, the parallelogram mechanism gravity G_{2i} ($i = 1, 2, 3$) and the gravity of A_2, G_{3i} ($i = 1, 2, 3$). The input torques are denoted as (M_1, M_2, M_3). The force which the operator's hand act on the hand-controller is denoted as $F = (F_x, F_y, F_z)$. Figure 5.12 shows the static mechanics model of the single kinematic chain.

According to the virtual work principle, there is Eq. (5.13):

$$M_1 \delta\theta_1 + M_2 \delta\theta_2 + M_3 \delta\theta_3 + F^T \delta X + \delta P = 0 \qquad (5.13)$$

where $\delta\theta_1, \delta\theta_2, \delta\theta_3$ are the virtual angular displacement under the driving torques; δX, ($\delta X = (\delta x, \delta y, \delta z)^T$) are the virtual

Figure 5.12 The static mechanics model of the single kinematic chain.

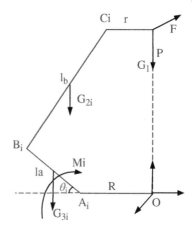

displacement under the application of F. P is the potential energy of this hand-controller.

$$P = G_1 z_1 + \sum_{i=1}^{3} G_{2i} z_{2i} + \sum_{i=1}^{3} G_{3i} z_{3i} \tag{5.14}$$

$$\delta P = G_1 \delta z_1 + \sum_{i=1}^{3} G_{2i} \delta z_{2i} + \sum_{i=1}^{3} G_{3i} \delta z_{3i} \tag{5.15}$$

where δz_{2i} and δz_{3i} denote the virtual displacements along the z-axis of the parallelogram mechanism of the No. i kinematic chain and the A_2 link, respectively.

So, substituting Eq. (5.15) into Eq. (5.13), we get Eq. (5.16):

$$M_1 \delta \theta_1 + M_2 \delta \theta_2 + M_3 \delta \theta_3 + G_1 \delta z_1 + \sum_{i=1}^{3} G_{2i} \delta z_{2i}$$

$$+ \sum_{i=1}^{3} G_{3i} \delta z_{3i} + F^T \delta X = 0 \tag{5.16}$$

According to the static mechanics model of the single kinematic chain shown in Figure 5.12, the z-axis coordinates of each moving part can be solved as:

$$\begin{cases} z_1 = P_z \\ z_{2i} = \dfrac{P_z + l_a \sin \theta_i}{2} \\ z_{3i} = \dfrac{l_a}{2} \sin \theta_i \end{cases} \tag{5.17}$$

So, the virtual displacements along the z-axis of these moving parts can be calculated as:

$$\begin{cases} \delta z_1 = \delta P_z \\ \delta z_{2i} = \dfrac{\delta P_z + l_a \delta \theta_i \cos \theta_i}{2} \\ \delta z_{3i} = \dfrac{l_a}{2} \delta \theta_i \cos \theta_i \end{cases} \tag{5.18}$$

δz is the virtual displacement of G_1 along the z-axis.

So, Eq. (5.16) can be transformed as:

$$M_1 \delta \theta_1 + M_2 \delta \theta_2 + M_3 \delta \theta_3 + G_1 \delta z + \sum_{i=1}^{3} G_{2i} \frac{\delta z + l_a \delta \theta_i \cos \theta_i}{2}$$

$$+ \sum_{i=1}^{3} G_{3i} \frac{l_a}{2} \delta \theta_i \cos \theta_i + F^T \delta X = 0 \tag{5.19}$$

where $F^T \delta X = F_x \delta x + F_y \delta y + F_z \delta z$.

According to the arbitrariness principle of virtual displacement, the condition that the mechanism maintains equilibrium at a certain position is that the sum of the virtual work made by any active force acting on the mechanism is equal to zero in any virtual displacement. So, setting $(\delta \theta_2, \delta \theta_3)$ to be $(0, 0)$, Eq. (5.18) is derived:

$$M_1 = - \left(G_1 + \frac{G_{21} + G_{22} + G_{23}}{2} + F_z \right) \frac{\partial z}{\partial \theta_1}$$

$$- \frac{G_{21} + G_{31}}{2} l_a \cos \theta_1 - F_x \frac{\partial x}{\partial \theta_1} - F_y \frac{\partial y}{\partial \theta_1} \tag{5.20}$$

Likewise, setting $(\delta \theta_1, \delta \theta_3)$ and $(\delta \theta_1, \delta \theta_2)$ to each be $(0, 0)$, Eqs. (5.21) and (5.22) are derived:

$$M_2 = - \left(G_1 + \frac{G_{21} + G_{22} + G_{23}}{2} + F_z \right) \frac{\partial z}{\partial \theta_2}$$

$$- \frac{G_{22} + G_{32}}{2} l_a \cos \theta_2 - F_x \frac{\partial x}{\partial \theta_2} - F_y \frac{\partial y}{\partial \theta_2} \tag{5.21}$$

$$M_3 = - \left(G_1 + \frac{G_{21} + G_{22} + G_{23}}{2} + F_z \right) \frac{\partial z}{\partial \theta_3}$$

$$- \frac{G_{23} + G_{33}}{2} l_a \cos \theta_3 - F_x \frac{\partial x}{\partial \theta_3} - F_y \frac{\partial y}{\partial \theta_3} \tag{5.22}$$

These equations include items of partial derivatives. $\frac{\partial x}{\partial \theta_i}, \frac{\partial y}{\partial \theta_i}, \frac{\partial z}{\partial \theta_i} (i = 1, 2, 3)$, should be solved.

From the kinematic analysis of the relationship between the coordinates of P, (x, y, z), and angular displacement of each motor, $(\theta_1, \theta_2, \theta_3)$ can be obtained in Eq. (5.23):

$$x^2 + y^2 + z^2 - 2x\cos\eta_i(\Delta r + l_a \cos\theta_i) - 2y \sin\eta_i(\Delta r + l_a \cos\theta_i)$$
$$-2zl_a \sin\theta_i + \Delta r^2 + 2\Delta r l_a \cos\theta_i + l_a^2 - l_b^2 = 0 \qquad (5.23)$$

Let

$$F(x, y, z, \theta_i) = x^2 + y^2 + z^2 - 2x\cos\eta_i(\Delta r + l_a \cos\theta_i) - 2y \sin\eta_i$$
$$\left(\Delta r + l_a \cos\theta_i\right) - 2zl_a \sin\theta_i + \Delta r^2 + 2\Delta r l_a \cos\theta_i + l_a^2 - l_b^2 \qquad (5.24)$$

According to $\frac{\partial x}{\partial \theta_i} = \frac{\partial F}{\partial \theta_i} / \frac{\partial F}{\partial x}, \frac{\partial y}{\partial \theta_i} = \frac{\partial F}{\partial \theta_i} / \frac{\partial F}{\partial y}, \frac{\partial z}{\partial \theta_i} = \frac{\partial F}{\partial \theta_i} / \frac{\partial F}{\partial z}$, and Eq. (5.24), Eq. (5.25) can be derived:

$$\begin{cases} \dfrac{\partial F}{\partial x} = 2x - 2\cos\eta_i(\Delta r + l_a \cos\theta_i) \\[2mm] \dfrac{\partial F}{\partial y} = 2y - 2\sin\eta_i(\Delta r + l_a \cos\theta_i) \\[2mm] \dfrac{\partial F}{\partial z} = 2z - 2l_a \sin\theta_i \\[2mm] \dfrac{\partial F}{\partial \theta_i} = \sin\theta_i(2x\cos\eta_i l_a + 2y\sin\eta_i l_a - 2\Delta r l_a) - 2zl_a \cos\theta_i \end{cases} \qquad (5.25)$$

So,

$$\begin{cases} \dfrac{\partial x}{\partial \theta_i} = \dfrac{\sin\theta_i(x\cos\eta_i l_a + y\sin\eta_i l_a - \Delta r l_a) - zl_a \cos\theta_i}{x - \cos\eta_i(\Delta r + l_a \cos\theta_i)} \\[3mm] \dfrac{\partial y}{\partial \theta_i} = \dfrac{\sin\theta_i(x\cos\eta_i l_a + y\sin\eta_i l_a - \Delta r l_a) - zl_a \cos\theta_i}{y - \sin\eta_i(\Delta r + l_a \cos\theta_i)} \\[3mm] \dfrac{\partial z}{\partial \theta_i} = \dfrac{\sin\theta_i(x\cos\eta_i l_a + y\sin\eta_i l_a - \Delta r l_a) - zl_a \cos\theta_i}{z - l_a \sin\theta_i} \end{cases} \qquad (5.26)$$

So substituting Eq. (5.26) into Eqs. (5.18), (5.21), and (5.22), the driving torque of each motor can be solved given the feedback force (F_x, F_y, F_z) and $(\theta_1, \theta_2, \theta_3)$, which are measured by encoders.

5.2.3 Friction Compensation

The desirable characteristics of a haptic display device include low friction and low inertia. However, most mechanical haptic devices tend to have some amount of friction. If a haptic display requires multi-DoF, multiple joints with friction is unavoidable [7]. Friction is a complex nonlinear phenomenon that has many negative effects on control systems, such as stiction, hysteresis, Stribeck effect, stick–slip, velocity dependence, and input frequency dependence. These become particularly significant when the motion is at low velocity, especially near zero crossings.

As we know, friction exists in the mechanical components of a haptic device, and it is highly nonlinear. To achieve high performance in the haptic display, modeling and compensation of the friction effects become necessary. Traditionally, friction is identified via a model-based approach.

For the serial haptic device Phantom Premium 1.5, its dynamics model can be formulated as shown in Eq. (5.27) [8].

$$\tau = M(\Theta)\ddot{\Theta} + C(\Theta, \dot{\Theta})\dot{\Theta} + N(\Theta) \tag{5.27}$$

where M, C, and N represent the inertial matrix, the Coriolis and centrifugal matrix, and the gravitational vector, respectively, defined in terms of the inertial and kinematic properties of the individual components of this haptic device. Here,

$$\tau = [\tau_1 \tau_2 \tau_3]^T \text{ and } \Theta = [\theta_1 \ \theta_2 \ \theta_3]^T$$

are the torque vectors delivered by the motors and the vector of joint angles derived from the encoders, respectively.

Using the Euler-Lagrange method, Cavusoglu et al. derived the dynamics structure and equations of motion for this haptic device as given in Eq. (5.28) [8].

$$
\begin{bmatrix} \tau_1 \\ \tau_2 \\ \tau_3 \end{bmatrix} = \begin{bmatrix} M_{11} & 0 & 0 \\ 0 & M_{22} & M_{23} \\ 0 & M_{32} & M_{33} \end{bmatrix} \begin{bmatrix} \ddot{\theta}_1 \\ \ddot{\theta}_2 \\ \ddot{\theta}_3 \end{bmatrix}
$$
$$
+ \begin{bmatrix} C_{11} & C_{12} & C_{13} \\ C_{21} & 0 & C_{23} \\ C_{31} & C_{32} & 0 \end{bmatrix} \begin{bmatrix} \dot{\theta}_1 \\ \dot{\theta}_2 \\ \dot{\theta}_3 \end{bmatrix} + \begin{bmatrix} 0 \\ N_2 \\ N_3 \end{bmatrix} \tag{5.28}
$$

Eq. (5.28) can further be transformed into Eq. (5.29) [9].

$$\tau = Y(\Theta, \dot{\Theta}, \ddot{\Theta})\pi \tag{5.29}$$

where Y is the regressor matrix and is π is the vector of eight unknown parameters, as defined in Eqs. (5.30) and (5.31).

$$
Y^T = \begin{bmatrix} Y_d^T \\ \hline Y_g^T \end{bmatrix}
$$

$$
= \begin{bmatrix}
\ddot{\theta}_1 & 0 & 0 \\[4pt]
\begin{matrix}\ddot{\theta}_1 c_{2.2} - 2\dot{\theta}_1\dot{\theta}_2 s_2 c_2 \\ -\dot{\theta}_1\dot{\theta}_2 s_{2.2}\end{matrix} & \dot{\theta}_1^2 s_{2.2} & 0 \\[10pt]
\begin{matrix}\ddot{\theta}_1 c_{2.3} - 2\dot{\theta}_1\dot{\theta}_3 s_3 c_3 \\ -\dot{\theta}_1\dot{\theta}_3 s_{2.3}\end{matrix} & 0 & \dot{\theta}_1^2 s_{2.3} \\[10pt]
\begin{matrix}\ddot{\theta}_1 c_2 s_3 - \dot{\theta}_1\dot{\theta}_2 s_2 s_3 \\ -\dot{\theta}_1\dot{\theta}_3 c_2 c_3\end{matrix} & \begin{matrix}-\tfrac{1}{2}\ddot{\theta}_3 s_{23} + \tfrac{1}{2}\dot{\theta}_1^2 s_2 s_3 \\ +\tfrac{1}{2}\dot{\theta}_3^2 c_{23}\end{matrix} & \begin{matrix}-\tfrac{1}{2}\dot{\theta}_2 s_{23} \\ -\tfrac{1}{2}\dot{\theta}_1^2 c_2 c_{23} \\ +\tfrac{1}{2}\dot{\theta}_2^2 c_{23}\end{matrix} \\[14pt]
0 & \ddot{\theta}_2 & 0 \\[4pt]
0 & 0 & \ddot{\theta}_3 \\[4pt]
\cdots\cdots & \cdots\cdots & \cdots\cdots \\[4pt]
0 & c_2 & 0 \\[4pt]
0 & 0 & s_3
\end{bmatrix}
\tag{5.30}
$$

$$
\pi = \begin{bmatrix} \pi_d \\ \hline \pi_g \end{bmatrix}
$$

$$
\begin{bmatrix}
\tfrac{1}{8}(4I_{ayy} + 4I_{azz} + 8I_{baseyy} + 4I_{beyy} + 4I_{bezz} + 4I_{cyy} + 4I_{czz} \\
\quad + 4I_{dfyy} + 4I_{dfzz} + L_1^2 m_c + L_2^2 m_a + 4L_3^2 m_c + 4L_1^2 m_a) \\[4pt]
\tfrac{1}{8}(4I_{beyy} - 4I_{bezz} + 4I_{cyy} - 4I_{czz} + 4L_1^2 m_a + L_1^2 m_c) \\[4pt]
\tfrac{1}{8}(4I_{ayy} - 4I_{azz} + 4I_{dfyy} - 4I_{dfzz} - L_2^2 m_a - 4L_3^2 m_c) \\[4pt]
L_1(L_2 m_a + L_3 m_c) \\[4pt]
\tfrac{1}{4}(4I_{bexx} + 4I_{cxx} + 4L_1^2 m_a + L_1^2 m_c) \\[4pt]
\tfrac{1}{4}(4I_{axx} + 4I_{dfxx} + L_2^2 m_a + 4L_3^2 m_c) \\[4pt]
\hline
g/2(2L_1 m_a + 2L_5 m_{be} + L_1 m_c) \\[4pt]
g/2(L_2 m_a + 2L_3 m_c - 2L_6 m_{df})
\end{bmatrix}
\tag{5.31}
$$

where π_d and π_g represent dynamic and gravitational parameter vectors, respectively.

The inertial and kinematic parameters $L_1, L_2, L_3, L_5, L_6, I_{axx}, I_{ayy}, I_{azz}, I_{azz}, I_{cxx}, I_{cyy}, I_{czz}, I_{baseyy}, I_{bexx}, I_{beyy}, I_{bezz}, I_{dfxx}, I_{dfyy}, I_{dfzz}, m_a, m_c, m_{be}$, and

m_{df} are the same as the ones defined previously [8], and g is the gravity acceleration.

The Y matrix contains all the terms in Eq. (5.27) that are the functions of the haptic device configuration vector Θ.

Here, s_i, c_i, s_{ij}, c_{ij}, $s_{2,i}$, and $c_{2,i} i, j = 1, 2, 3$, represent the shorthand notation for $\sin(\theta_i)$, $\cos(\theta_i)$, $\sin(\theta_i - \theta_j)$, $\cos(\theta_i - \theta_j)$, $\sin(2\theta_i)$, and $\cos(2\theta_i)$, respectively.

The following Coulomb and viscous friction models

$$\tau_{fi} = \pi_{fci}\, sgn(\dot{\theta}_i) + \pi_{fvi}\dot{\theta}_i, \quad i = 1, 2, 3$$

or

$$\tau_f = \begin{bmatrix} \tau_{f1} \\ \tau_{f2} \\ \tau_{f3} \end{bmatrix} = \begin{bmatrix} \pi_{fc_1}\, \mathrm{sgn}(\dot{\theta}_1) + \pi_{fv_1}\dot{\theta}_1 \\ \pi_{fc_2}\, \mathrm{sgn}(\dot{\theta}_2) + \pi_{fv_2}\dot{\theta}_2 \\ \pi_{fc_3}\, \mathrm{sgn}(\dot{\theta}_3) + \pi_{fv_3}\dot{\theta}_3 \end{bmatrix} = \mathbf{Y}_f\pi_f$$

are also employed to include the effect of friction, where sgn(.) denotes the signum function, π_{fc_i} and π_{fv_i}, represents the Coulomb and viscous friction coefficients for joint i, respectively, and Y_f and π_f are defined as:

$$\mathbf{Y}_f = \begin{bmatrix} \dot{\theta}_1 & 0 & 0 & sgn(\dot{\theta}_1) & 0 & 0 \\ 0 & \dot{\theta}_2 & 0 & 0 & sgn(\dot{\theta}_2) & 0 \\ 0 & 0 & \dot{\theta}_3 & 0 & 0 & sgn(\dot{\theta}_3) \end{bmatrix}$$

$$\pi_f = \begin{bmatrix} \pi_{fv1} & \pi_{fv2} & \pi_{fv3} & \pi_{fc1} & \pi_{fc2} & \pi_{fc3} \end{bmatrix}^T$$

Considering the friction effects in the dynamic models, the total Y and π expand to:

$$\pi = \begin{bmatrix} \pi_d \\ -- \\ \pi_g \\ -- \\ \pi_f \end{bmatrix}$$

and

$$Y_{[3 \times 14]} = [Y_{d[3 \times 6]} \quad Y_{g[3 \times 2]} \quad Y_{f[3 \times 6]}]$$

The linear system of equations in Eq. (5.29) can be solved using the least squares estimation method if several independent data points are available. To collect this data, the haptic device is moved along a trajectory and its joint angles and motor torques are recorded for a period of time to create π_N and Y_N, which are an ensemble of the torque vectors and regression matrices τ, and Y stacked over for N samples. The least squares solution for π is:

$$\hat{\pi} = \left[(Y_N^T Y_N)^{-1} Y_N^T \right] \tau_N$$

where $\hat{\pi}$ is the estimate of the π vector, that minimizes the torque error in the sense of mean square, and $(Y_N^T Y_N)^{-1} Y_N^T$ is called the left pseudo-inverse of Y_N. The number of measurements should be high enough to avoid ill-conditioning of matrix Y_N and to assure the existence of the pseudo-inverse matrix.

Experimental parameter identification cannot determine the property of each dynamic component of this haptic device. Specifically, Eq. (5.31) illustrates that the first eight elements of the π vector are each combination of several mass, inertial, and length properties for various links of this hand-controller. It should be noted that while the properties of this manioulandum and components are not individually identified, the set of identified parameters is sufficient for model-based control and hand force estimation.

5.3 Calculation of Virtual Force

In a virtual environment, the interaction forces depend on the geometry of the object being touched, its compliance, and the geometry of the avatar representing the haptic interface.

Haptic rendering can be split into three main blocks, as shown in Figure 5.13. Collision-detection algorithms provide information about contact occurring between an avatar at position X and objects in the virtual environment [2]. Force-response algorithms return the ideal interaction force, F_d, between an avatar and virtual objects. Control algorithms return a force F_r to the user, approximating the ideal interaction force to the best of the device's capabilities.

5.3.1 Collision Detection

Collision detection algorithms detect collisions between objects and avatars in the virtual environment and yield information about where and when collision have occurred, and, ideally, the extent of the collisions (e.g., penetrations, indentations, contact area, etc.). The problems related to collision detection and distance computations have been extensively studied in computational geometry, robotics, simulated environments, and haptics [10–19].

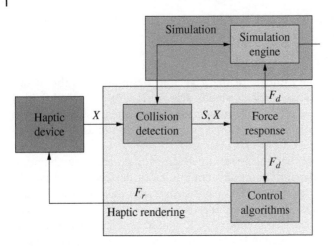

Figure 5.13 Three main blocks of haptic rendering.

In collision detection, improving the calculation accuracy of the algorithm involves a tradeoff with the calculation latency. If the accuracy is preferred, the calculation time of the algorithm and the complexity will be greatly increased, thereby increasing the latency. On the contrary, if the calculation needs to be in real-time, the accuracy of the algorithm will be sacrificed to a certain extent. Existing collision detection algorithms are based on the specific accuracy and real-time needs.

From the time perspective, collision detection algorithms can be divided into static collision detection and dynamic collision detection [20]. Static collision detection is applied when the scene does not change over time. Dynamic collision detection is applied to varying scenarios. The dynamic collision detection can be subdivided into discrete collision detection and continuous collision detection [21], depending on whether the detection time is discrete or continuous.

From the perspective of the spatial domain, collision detection is either based on solid space or image space. Collision detection based on solid space can be categorized based on the types of models: convex polytopes and general polygonal models. For convex polytopes, various techniques have been developed based on linear programming [22], incremental computation of Minkowski difference [23, 24], feature tracking based on Voronoi regions [14, 25] and multi-resolution methods [26, 27]. Some of these algorithms are based on incremental computations and exploit frame-to-frame coherence [14, 23, 25].

For general polygonal models, bounding volume hierarchies (BVHs) have been widely used for collision detection and separation distance queries [28]. Hierarchies differ based on the underlying bounding volume or transversal schemes. These include the AABB trees [29], OBB trees [30, 31], sphere trees [32], k-DOPs [33], swept sphere volumes [34], and convex hull-based trees [35]. Some of these are shown in Figure 5.14.

Next, a new kind of bounding box based on the center axis is introduced, which is used in a manipulator. It simplifies the geometrical models and speed up the algorithm.

5.3.1.1 The Construction of the Bounding Box

For a three-dimensional model, the surface can be divided into several triangular patches. Assume that there are n triangular patches. The three vertices of the No. i triangular patch are denoted as p_i, q_i, r_i, which are 3×1 vectors.

The axis-aligned bounding box is one of the most widely used bounding boxes. The axial bounding box is six-sided, box-like, and each face normal is parallel to a coordinate axis. The axis-aligned bounding box can achieve quick intersection testing, but its wrapping ability is poor. When the object rotates, it needs to constantly update the axis-aligned bounding box.

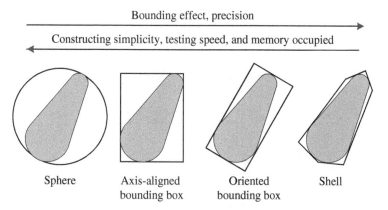

Bounding effect, precision

Constructing simplicity, testing speed, and memory occupied

| Sphere | Axis-aligned bounding box | Oriented bounding box | Shell |

Figure 5.14 2D schematic diagram of several common bounding boxes.

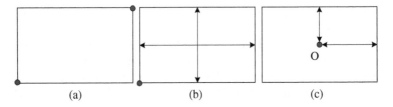

Figure 5.15 Three presenting methods of axis-aligned bounding box.

There are three presenting methods of the axis-aligned bounding box shown in Figure 5.15.

In the first method (Figure 5.15a), the maximum and minimum values along each axis are used to define the box. These values are obtained by comparing coordinates of the vertices of triangular patches of a three-dimensional object in the world coordinate system.

In the second method (Figure 5.15b), the vertices with the minimum value along each axis are used together with the length ranges d_x, d_y, and d_z on these axes. These minimum values are also obtained by comparing coordinates of the vertices of triangular patches of a three-dimensional object in the world coordinate system. d_x, d_y, and d_z are calculated as below:

$$d_x = \max\{p_i(x), q_i(x), r_i(x)\} - \min\{p_i(x), q_i(x), r_i(x)\}$$

$$d_y = \max\{p_i(y), q_i(y), r_i(y)\} - \min\{p_i(y), q_i(y), r_i(y)\}$$

$$d_z = \max\{p_i(z), q_i(z), r_i(z)\} - \min\{p_i(z), q_i(z), r_i(z)\}$$

In the third method (Figure 5.15c), the center of the bounding box, O, and halves of length ranges along each axis, r_x, r_y, and r_z are used. The x, y, and z coordinates of center O are:

$$x = \frac{\max\{p_i(x), q_i(x), r_i(x)\} - \min\{p_i(x), q_i(x), r_i(x)\}}{2}$$

$$y = \frac{\max\{p_i(y), q_i(y), r_i(y)\} - \min\{p_i(y), q_i(y), r_i(y)\}}{2}$$

$$z = \frac{\max\{p_i(z), q_i(z), r_i(z)\} - \min\{p_i(z), q_i(z), r_i(z)\}}{2}$$

In the construction of the axis alignment bounding box tree, the top-down approach is generally used. The child node divides the geometric objects of the parent node, and then builds the axis-aligned bounding box for the divided geometric structure, then reiterate to divide geometric structures until the wrapping requirements are met.

The intersection test of these axis alignment bounding boxes uses the interval test method. These axis alignment bounding boxes are projected onto the three axes, respectively. If these projected intervals of two axis-aligned bounding boxes overlap each other on all three axes, it is known that the two bounding boxes intersect, i.e. collide. Otherwise, as long as there is no overlapping of projected intervals on one axis, it is determined that these two bounding boxes does not intersect.

For a robot arm model, for the consideration of the motion stability, each mechanical link of the robot arm will generally have a central axis, the link model around the central axis is symmetrical. In most cases, the central axis is a straight-line segment, in a few cases, the central axis is a polyline with all angles being right, which can be divided into several straight lines to consider. Here, center axis–based bounding box is illustrated.

The three-dimensional model and schematic diagram of a 7-DoF robot arm mechanism are shown in Figure 5.16.

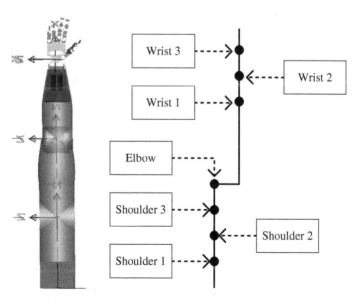

Figure 5.16 3D model of a robot arm (left) and schematic diagram of mechanism (right).

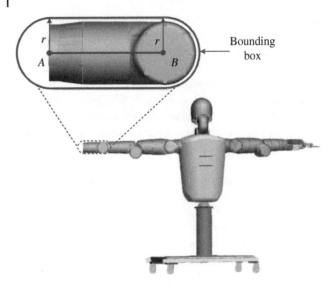

Figure 5.17 Schematic diagram of bounding box.

Since the robot arm link model is symmetrical about the central axis, a cylindrical bounding box with the central axis of the arm linkage model can be used instead of the robotic link. At the same time, for the purposes of anti-collision and in order to ensure the distance detection algorithm is simple and efficient, the bounding box can be properly expanded at both ends of the cylindrical bounding box with two hemispherical bounding boxes. Taking the joint with the straight central axis as an example, the final bounding box is shown in Figure 5.17.

Here, the bounding box is represented by the straight line, *AB*, representing the central axis and *r* the radius of the bounding box. Generally, point A and point B are special points on the robot arm (Figure 5.18). Their coordinates can be calculated using robot kinematic equations. The radius is measured from its three-dimensional models. For an object in a virtual scene other than the robot arm, one or several spherical bounding boxes are used.

5.3.1.2 Calculation of Distance between Bounding Boxes

Once the bounding boxes are built, distances between these bounding boxes must be calculated. Based on these distances, it can be checked whether these bounding boxes intersect, i.e., whether related objects collide. This problem can be transformed into a simple spatial geometry problem – that is, to solve the shortest distance from one

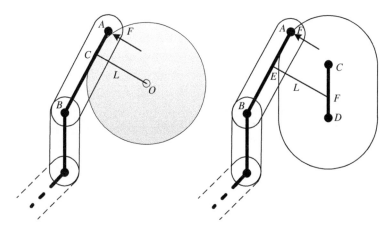

Figure 5.18 Schematic diagram of distance calculation.

point or one straight segment to another straight segment, as shown in Figure 5.18.

Bounding Box Is a Sphere

In the case of the bounding box as a sphere, the distance between the center of the bounding sphere, O, and the center axis, AB, and the closest point, C, on the straight line AB from the center O is solved. That is to make a perpendicular from the center, O, to the straight line AB and calculate the its length L, and seek the perpendicular foot, C.

Because $\overrightarrow{OC} \perp \overrightarrow{AB}$ and C is on AB, first seek the plane Ω, which is perpendicular to $\overrightarrow{AB}$ and passes through point O. Its equation is denoted as:

$$(x - x_O)(x_B - x_A) + (y - y_O)(y_B - y_A) + (z - z_O)(z_B - z_A) = 0 \tag{5.32}$$

The straight line AB can be denoted as:

$$\frac{x - x_A}{x_B - x_A} = \frac{y - y_A}{y_B - y_A} = \frac{z - z_A}{z_B - z_A} = t \tag{5.33}$$

So,

$$\begin{cases} x = x_A + t(x_B - x_A) \\ y = y_A + t(y_B - y_A) \\ z = z_A + t(z_B - z_A) \end{cases} \tag{5.34}$$

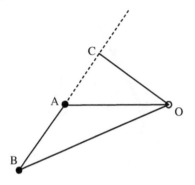

Figure 5.19 Perpendicular foot is not on the central axis.

From Eqs. (5.32) and (5.34), (x_C, y_C, z_C), the coordinates of perpendicular foot C, can be determined, and the shortest distance L is:

$$L = \sqrt{(x_O - x_C)^2 + (y_O - y_C)^2 + (z_O - z_C)^2} \tag{5.35}$$

Since the central axis AB is a straight segment, not a straight line, the perpendicular foot C may not be on the straight segment AB, as shown in Figure 5.19.

So the shortest distance L is set as:

$$L = \min\{|OA|, |OB|\} \tag{5.36}$$

Bounding Box Is Center Axis–Based Bounding Box

If the bounding box is based on the center axis, the shortest distance between two straight segments, AB and CD, must be calculated. Let (x_A, y_A, z_A) be the coordinates of A, (x_B, y_B, z_B), be the coordinates of B. (x_C, y_C, z_C), be the coordinates of C, and (x_D, y_D, z_D) be the coordinates of D.

Take an arbitrary point E on AB. Let (x_E, y_E, z_E) be the coordinates of E. So:

$$\begin{cases} x_E = x_A + t(x_B - x_A) \\ y_E = y_A + t(y_B - y_A) \\ z_E = z_A + t(z_B - z_A) \end{cases} \tag{5.37}$$

In Eq. (5.37), t is the only argument. When $0 < t < 1$, point E lies on the straight segment AB. When $t > 1$, point E lies on the extension of the straight segment AB; When $t < 0$, point E lies on the extension of the straight segment BA.

Likewise, take an arbitrary point F on the straight segment CD. Let (x_F, y_F, z_F) be the coordinates of point F. So:

$$\begin{cases} x_F = x_C + s(x_D - x_C) \\ y_F = y_C + s(y_D - y_C) \\ z_F = z_C + s(z_D - z_C) \end{cases} \quad (5.38)$$

Then, the distance between E and F is solved as:

$$|EF| = \sqrt{(x_E - x_F)^2 + (y_E - y_F)^2 + (z_E - z_F)^2} \quad (5.39)$$

Let $f(s, t) = |EF|^2$, so,

$$\begin{aligned} f(s, t) &= (x_E - x_F)^2 + (y_E - y_F)^2 + (z_E - z_F)^2 \\ &= [(x_A - x_C) + t(x_B - x_A) - s(x_D - x_C)]^2 \\ &\quad + [(y_A - y_C) + t(y_B - y_A) - s(y_D - y_C)]^2 \\ &\quad + [(z_A - z_C) + t(z_B - z_A) - s(z_D - z_C)]^2 \end{aligned} \quad (5.40)$$

To find the shortest distance between AB and CD, the minimum value of $f(s, t)$ is determined.

Find the partial derivative of $f(t, s)$ with respect to t and s, and set it equal to zero, so:

$$\begin{cases} \dfrac{\partial f(t, s)}{\partial t} = 0 \\[2mm] \dfrac{\partial f(t, s)}{\partial s} = 0 \end{cases} \quad (5.41)$$

Solving the group of Eqs. (5.41) yields the values of t and s.

If $0 < t < 1$ and $0 < s < 1$, point E lies on the straight segment AB, and point F lies on the straight segment CD. The shortest distance L between AB and CD is the distance between E and F:

$$L = |EF| = \sqrt{(x_E - x_F)^2 + (y_E - y_F)^2 + (z_E - z_F)^2} \quad (5.42)$$

If $0 < s < 1$, $t < 0$, or $t > 1$, then point E is not on the straight segment AB and point F lies on the straight segment CD. So, the closest point on the straight line AB from the straight segment CD is not on the straight segment AB. The distances L_1, L_2 of point A and point B from the straight segment CD need to be calculated. Then, the shortest distance L between the straight segments AB and CD is:

$$L = \min\{L_1, L_2\} \quad (5.43)$$

If $0 < t < 1$, $s < 0$, or $s > 1$, point E lies on the straight segment AB, and point F does not lie on the straight segment CD. So, the shortest distances need to be calculated from point C and point D to the straight segment AB, L_3, L_3, respectively. Then, the shortest distance L between straight segments, AB, CD is:

$$L = \min\{L_3, L_4\} \tag{5.44}$$

If $t < 0$ or $t > 1$, and $s < 0$, or $s > 1$, point E is not on the straight segment AB and point F is not on the straight segment CD. Then, the shortest distances need to be calculated from point A and point B to the straight segment CD, L_1, L_2, respectively, along with the shortest distance from point C and point D to the straight segment AB. So, the shortest distance between straight segments AB and CD, L is:

$$L = \min\{L_1, L_2, L_3, L_4\} \tag{5.45}$$

If L is less than the two times the bounding box, it can be considered that collision occurs. Sometimes, some additional conditions are added for further safety consideration.

5.3.2 Calculating the Model of Virtual Force

After a collision is detected based on the calculation of distance between two bounding boxes, force-response algorithms compute the interaction force between these two virtual objects. This force approximates as closely as possible the contact forces that would normally arise during contact between real objects. Force-response algorithms typically operate on the positions of these virtual objects and the collision state between them. Their return values are normally force and torque vectors that are applied at the device-body interface.

5.3.2.1 1-DoF Interaction

A 1-DoF device measures the operator's position and applies forces to the operator along one spatial dimension only. Types of 1-DoF interactions include opening a door with a knob that is constrained to rotate around one axis, squeezing scissors to cut a piece of paper, or pressing a syringe's piston when injecting a liquid into a patient. A 1-DoF interaction might initially seem limited; however, it can render many interesting and useful effects.

Figure 5.20 An example of 1-DoF interaction: the virtual wall concept.

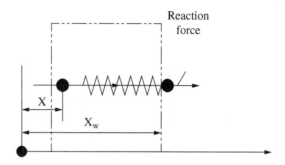

Rendering a virtual wall – that is, creating the interaction forces that would arise when contacting an infinitely stiff object – is the prototypical haptic task. As one of the most basic forms of haptic interaction, it often serves as a benchmark in studying haptic stability. The discrete-time nature of haptic interaction means that the haptic interface avatar will always penetrate any virtual object. A positive aspect of this is that the force-rendering algorithm can use information on how far the avatar has penetrated the object to compute the interaction force. As Figure 5.20 illustrates, if we assume the avatar moves along the x-axis and $x < x_w$ describes the wall, the simplest algorithm to render a virtual wall is given by [2]:

$$F = \begin{cases} 0 & x > x_w \\ K(x_w - x) & x \le x_w \end{cases} \tag{5.46}$$

where K represents the wall's stiffness and, thus, is ideally very large; F is the virtual force. More interesting effects can be accomplished for a 1-DoF interaction [7, 8].

5.3.2.2 2-DoF Interaction

To feel the contour of a planar figure, 2-DoF forces need to be calculated. Generally, the contour line is described in NURBS (Non-Uniform Rational B-Splines). Sometimes the contour line is generated by an image-processing method, given the image of that planar figure. On each point of that planar curve, the virtual force is calculated along the perpendicular direction, as shown in Figure 5.21. The planar reaction force is calculated by the spring model too. At different points, the direction of the reaction force is different. But it is always normal to the curve at the contact point. l here is the penetration depth of the avatar into the other side of the curve.

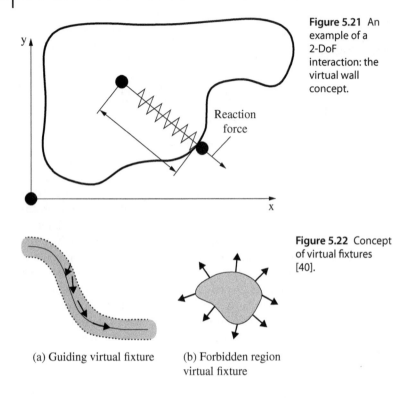

Figure 5.21 An example of a 2-DoF interaction: the virtual wall concept.

Reaction force

Figure 5.22 Concept of virtual fixtures [40].

(a) Guiding virtual fixture

(b) Forbidden region virtual fixture

5.3.2.3 3-DoF Interaction

In a 3-DoF interaction, we assume that we interact with the virtual world with a point probe, which requires that we only compute the three interaction force components at the probe's tip. This greatly simplifies the interface device design and facilitates collision detection and force computation [2].

The virtual fixture was presented by Rosenberg [36]. It is used to guide the operator's motion along a certain trajectory or path [37], or confine the motion inside certain parts of the workspace, or to prevent penetration into undesired parts of the workspace [36, 38, 39], as shown in Figure 5.22 [40].

When a virtual fixture is used in a teleoperation system, generally virtual force should be calculated and fed back to the haptic device to let the operator sense it. Artificial potential field, first presented by Khatib [41] is generally used to calculate the virtual force. If the virtual object considered is just simplified as a mass point, the virtual force acting on this point just has three dimensions. The artificial potential field theory was originally developed for an online collision avoidance system [42–44] so there must exist one path that can lead the robot toward the target and avoid obstacles automatically in the virtual potential field. Therefore, it is possible to develop the relationship between the virtual force and the potential function, which is the key part of the new force guidance strategy [45].

For a forbidden-region virtual fixture that surrounds the obstacle, the function of artificial potential field can be built as:

$$U_{rep} = \begin{cases} \dfrac{1}{2}\eta \left(\dfrac{1}{\rho} - \dfrac{1}{\rho_0} \right)^2 & \rho \le \rho_0 \\ 0 & \rho > \rho_0 \end{cases} \tag{5.47}$$

where ρ represents the distance between the obstacle and the end-effector of robot along its velocity direction, ρ_0 denotes the influence distance of this potential field, and η is a constant that determines the magnitude of repulsive potential.

The function U_{rep} is positive or null and reaches to a maximum at the boundary of the obstacle, but it will reduce to zero when the robot is sufficiently away from it. The repulsive force F_{rep} can be generated as

$$F_{rep} = -\nabla U_{rep} = \begin{cases} \eta \left(\dfrac{1}{\rho} - \dfrac{1}{\rho_0} \right) \dfrac{1}{\rho^2} \dfrac{\partial \rho}{\partial \mathbf{r}} & \rho \le \rho_0 \\ 0 & \rho > \rho_0 \end{cases} \tag{5.48}$$

where $\mathbf{r}$ is the position vector of the virtual end-effector of robot; $\dfrac{\partial \rho}{\partial \mathbf{r}} = \left[\dfrac{\partial \rho}{\partial x} \dfrac{\partial \rho}{\partial y} \dfrac{\partial \rho}{\partial z} \right]^T$.

In guiding virtual fixtures, the attractive potential field U_{att} is simply defined as:

$$U_{att} = 0.5\mu \|\mathbf{r} - \mathbf{r}'\|^2 \tag{5.49}$$

where μ is a positive scaling factor, $\mathbf{r}$ is the position vector of the virtual end-effector of the robot, and $\mathbf{r}'$ is the vector of the perpendicular point on the guiding curve from the endpoint of $\mathbf{r}$. The attractive force is the negative gradient direction of the potential function:

$$F_{att} = -\nabla U_{att} = -\mu(\mathbf{r} - \mathbf{r}') \tag{5.50}$$

5.3.2.4 6-DoF Interaction

Although the point interaction metaphor has proven to be surprisingly convincing and useful, it has limitations, such as, simulating the interaction between a tool tip and a virtual environment means we cannot apply torques through the contact [2]. So, haptic rendering of more than 3 DoF have been researched. Barbagli et al. developed an algorithm to simulate a 4-DoF interaction through soft-finger contact [46]. Basdogan et al. implemented a 5-DoF interaction, such as that which occurs between a line segment and a virtual object, to approximate contact between long tools and virtual environments [47]. Haptic rendering of 4 and 5 DoF can be considered as the simplified cases of haptic rendering of 6 DoF. Many research studies have developed algorithms providing for 6-DoF interaction forces [48–61].

Generally, we can still use Hooke's law to generate the contact force to reduce computing complexity:

$$F = kD_p \tag{5.51}$$

where k is the spring stiffness constant and D_p is the depth of penetration calculated by collision detection. Generally, F has three components along the x-, y- and z-axes, respectively. F is in the contact normal direction. If the contact normal direction does not pass through the mass center of the tool, then there will be torque acting on the tool. The torque T is calculated by

$$\mathbf{T} = \sum_i \mathbf{R}_i \times \mathbf{F}_i \tag{5.52}$$

where $\mathbf{F}_i$ is the contact force vector applied at the point p_i, and $\mathbf{R}_i$ is the radius vector from the center of mass to p_i [53]. Generally, $\mathbf{T}$ also has three components rotating about the x, y, and z axes, respectively.

5.4 Haptic Display Based on Point Haptic Device

In this chapter, only the point haptic device (manipulandum) is discussed, although the haptic device includes force feedback type and tactile feedback type. The proxy object interacts with the virtual environment in multiple ways, such as, sliding or rubbing on the surface of another object. The feedback force and torque on the control point will allow the operator to sense the texture of that object, generally combining the visual image attached to that object. Besides the hand-controller, which can display tactile based on force feedback, there are many other types of tactile display devices, using different methods, such as mechanical needles actuated by electromagnetic technologies [58, 59], shape memory alloys [60, 61], piezoelectric crystals [62], pneumatic systems, heat pump systems, electrorheological fluids [63], and vibrotactile [64–67].

5.4.1 Human Tactile Perception

In the actual surface of the object, there are tiny irregularities, and by rubbing one's fingers over the actual surface of the object, the surface texture of the object can be perceived [2]. When the fingers contact with the external objects and produce relative movement, the finger skin generates compression, stretching, and other mechanical deformation under pressure and lateral movement [68, 69], induces the mechanical stimulation sensor in the deep skin to produce the corresponding action potential, and the tactile information is transmitted to the cerebral cortex to identify the object shape, texture, and other physical characteristics [70, 71].

The tactile perception has several psychophysical dimensions, including hardness (hard/soft), friction (moist/dry, sticky/slippery), roughness (fine roughness (rough/smooth), macro roughness (uneven, relief)), temperature (warm/cold) [72]. The role of friction and surface texture in tactile perception was investigated in particular detail, because typically tactile exploration involves moving (at least) one finger over a textured surface [73].

When designing haptic devices and haptic rendering, consideration should be paid to the psychophysics and biomechanics involved in the perception of vibratory and frictional cues. Typical human factor data are shown in the Table 5.3.

5.4.2 Haptic Texture Display Methods

In the virtual environment, people try to simulate the friction between the surfaces, so that the operator can use the tactile device to perceive the surface texture of the virtual object. The mechanism of surface friction is very complex, but many researchers have established simple or complex models, which become the basis of tactile rendering algorithms [2, 47, 49, 74–80].

Table 5.3 Some human factor data related to tactile sense [81].

Item	Value or Range	Remark
Temporal resolution	800 Hz	the maximum frequency that can be felt by the human somatosensory system [100]
	30 ms	The minimum delay unnoticeable between the user motion and the force rendering [101]
Spatial resolution	50 μm	Considering a slow exploration speed of 40 mm/s and the previously-mentioned frequency limit of $f = 800$ Hz, we estimate that the smallest perceptible wavelength is on the order of $\lambda = \frac{v}{f} = 50$ μm [72, 102, 103]
	0.2 mm	The minimum feature size that can be detected in static touch [84, 104]
Force resolution	10^{-2}N	the smallest static force that a human can perceive [105]
	5×10^{-4}N	The smallest dynamic force [106, 107]
	0.5 N	The peak lateral force to explore the texture
Force dynamic range	10^3	

The tactile perception of texture is influenced by a wide set of physical properties, including roughness at multiple-length scales, skin-surface adhesion, and surface deformability [81, 82]. Each of these properties, along with the mechanics of the finger itself, contribute to frictional losses that relate to the perception of slipperiness and stickiness [83], as well as vibrations that related to perceived roughness [84]. A challenge that faces the designers of haptic interfaces is emulating a wide range of tactile experiences with control over a substantially reduced set of physical variables [81].

The reproduction of the friction force as a vibration correlated with the motion of the fingers allows the rendering of complex roughness profiles. However, because vibrotactile actuation is limited to frequencies higher than 20 Hz, the quasi-static content is not represented, which implies that the stickiness dimension of the texture cannot be controlled [85].

Some researchers use a stylus to slide along a path on the surface of a real object to measure pressure, friction, and displacement data, and attach these data onto the surface of the virtual object, or

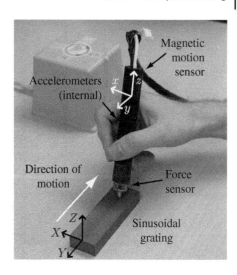

send to the haptic device directly, to allow the operator to feel the texture [86–90].

McDonald and Kuchenbecker [91] use a haptic recording device equipped with a suite of sensors to record a large number of interactions with the selected texture, as shown in Figure 5.23. They used this rigid probe to record interactions with a sinusoidal grating and built mathematical model to simulate the lateral and axial tool accelerations that were recorded. They used the Eq. (5.53) to calculate the normal forces, as drawn in Figure 5.24.

$$F_n = \frac{4}{3}E^* R_e^{\frac{1}{2}} \; (\delta/\mathscr{F}_2)^{\frac{3}{2}} + b_s(\delta/\mathscr{F}_2)^{\frac{3}{2}}\dot{\delta} \tag{5.53}$$

where E^* is defined by material properties, R_e is defined by the radii of curvature of the tool and surface, and $\mathscr{F}_2$ is a geometric correction factor based on the eccentricity of the predicted contact ellipse [91].

The exact solution for the tangential forces is

$$\boldsymbol{F}_t = -\mu(v)\boldsymbol{F}_n \tag{5.54}$$

where μ is the friction coefficient that is a function of the tangential sliding velocity between the two interacting bodies. This friction coefficient is calculated by the model:

$$\mu(v) = \mu_v v + \left[\mu_c + \sigma e^{-|v/v_c|} - (\sigma + \mu_c)e^{-|nv/v_c|}\right] \; \text{sign}(v)$$

$$\sigma = \frac{(2n^n)^{\frac{1}{n-1}}}{n-1}(\mu_s - \mu_c).$$

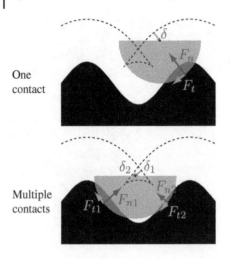

Figure 5.24 Calculation of penetration depth and resultant normal and tangential forces for two contact situations [107].

Some tactile rendering algorithms use a simple approach, such as calculating the bump data on the surface according to the desired texture (some calculate the depth of the points on the surface by visual processing algorithm, according to a bitmap attached to the surface [92–98]), and then the corresponding displacement of each contact point is directly generated on the haptic device [99], which generally is an array of vibration actuators using dynamic varying displacements of each actuator to generate the tactile feeling for the operator. For the force feedback device, the texture force and friction force are calculated based on the acquired height map [98]. The simplest method to calculate the contact force is finding the normal force is calculated using Hooke's law, where the normal force is proportional to the height at that position; the friction force is calculated according to the sliding velocity and the friction coefficient.

References

1 Lionel Birglen, Haptic Devices Based on Parallel Mechanisms. State of the Art, http://www.parallemic.org/Reviews/Review003.html.

2 Salisbury, K., Conti, F., and Barbagli, F. (2004). Haptic rendering: introductory concepts. *IEEE Comput. Graphics Appl.* 24 (2): 24–32.

3 Majid H. Koul, Praneeth Kumar, Praveen K. Singh, et al. Gravity Compensation for PHANToMTM Omni® Haptic Interface, The 1st Joint International Conference on Multibody System Dynamics May 25–27, 2010, Lappeenranta, Finland

4 Murat Cenk Cavusoglu, David Feygin, Kinematics and Dynamics of Phantom (TM) model 1.5 Haptic Interface, https://www2.eecs .berkeley.edu/Pubs/TechRpts/2001/ERL-01-15.pdf

5 Kinematics and Dynamics of Phantom(TM) model 1.5 Haptic Interface, Murat Cenk Cavusoglu, David Feygin, 2001

6 Tavakoli, M. (2008). *Haptics for Teleoperated Surgical Robotic System*. River Edge, NJ, USA: World Scientific Publishing Co., Inc. ISBN: 9812813152 9789812813152.

7 Bi, D., Li, Y.F., Tso, S.K. et al. (2004). Friction modeling and compensation for haptic display based on support vector machine. *IEEE Trans. Ind. Electron.* 51 (2): 491–500.

8 M. C. Cavusoglu and D. Feygin, "Kinematics and dynamics of phantom(tm) model 1.5 haptic interface," UC Berkeley ERL Memo M01/15, 2001. [Online]. Available: http://vorlon.cwru.edu.

9 Taati, B., Tahmasebi, A.M., and Hashtrudi-Zaad, K. (2008). Experimental identification and analysis of the dynamics of a PHANToM premium 1.5A haptic device. *Teleoperators Virtual Environ.* 17 (4): 327–343.

10 Young J. Kim, Miguel A. Otaduy, Ming C. Lin et al. Six-Degree-of-Freedom Haptic Display Using Localized Contact Computations, http://gamma.cs.unc.edu/6DOFLCC/haptic02 .pdf.

11 Yi-sheng, Z., Guo-fu, D., Ming-heng, X. et al. (2008). Survey on real-time collision detection algorithms. *Appl. Res. Comput.* 25 (1): 8–12.

12 Huard, B., Grossard, M., Moreau, S. et al. (2013). Position estimation and object collision detection of a tendon-driven actuator based on a polytopic observer synthesis. *Control Eng. Pract.* 21 (9): 1178–1187.

13 Lin M, Gottschalk S. Collision detection between geometric models: a survey 1998:602–608.

14 Mirtich, B. (1998). V-clip: fast and robust polyhedral collision detection. *ACM Trans. Graphics* 17 (3): 177–208.

15 Gilbert, E.G., Johnson, D.W., and Keerthi, S.S. (1988). A fast procedure for computing the distance between complex objects in three-dimensional space. *IEEE J. Robot. Autom.* 4 (2): 193–203.

16 Luca, A., Albu-Schaffer, A., and Haddadin, S. (2006). Collision detection and safe reaction with the DLR-III lightweight manipulator arm. *IEEE Intell. Robots Syst.* 1623–1630.

17 Turk, G. (1989). *Interactive Collision Detection for Molecular Graphics*. Chapel Hill: The University of North Carolina.

18 Teschner, M., Heidelberger, B., and Müller, M. (2003). Optimized spatial hashing for collision detection of deformable objects. *VMV* 3: 47–54.

19 Haddadin, S., Albu, A., Luca, A. et al. (2008). Collision detection and reaction: a contribution to safe physical human-robot interaction. *Intell. Robots Syst.* 3356–3363.

20 Chao Wang. Research on Collision Detection in the Virtual Assembly. Dissertation of East China University of Science and Technology, 2013

21 Tang, M., Curtis, S., Yoon, S. et al. (2009). ICCD: interactive continuous collision detection between deformable models using connectivity-based culling. *IEEE Trans. Visual Comput. Graphics* 15 (4): 544–557.

22 R. Seidel. Linear programming and convex hulls made easy. In Proc. 6th Ann. ACM Conf. on Computational Geometry, pages 211–215, Berkeley, California, 1990.

23 S. Cameron. Enhancing gjk: Computing minimum and penetration distance between convex polyhedra. Proceedings of International Conference on Robotics and Automation, pages 3112–3117, 1997.

24 Gilbert, E.G., Johnson, D.W., and Keerthi, S.S. (1988). A fast procedure for computing the distance between objects in three-dimensional space. *IEEE J. Robot. Autom.* RA-4: 193–203.

25 M.C. Lin and John F. Canny. Efficient algorithms for incremental distance computation. In IEEE Conference on Robotics and Automation, pages 1008–1014, 1991

26 S. Ehmann and M. C. Lin. Accelerated proximity queries between convex polyhedral using multi-level voronoi marching. Proc. of IROS, 2000.

27 L. Guibas, D. Hsu, and L. Zhang. H-Walk: Hierarchical distance computation for moving convex bodies. Proc. of ACM Symposium on Computational Geometry, 1999.

28 Clark, J.H. (1976). Hierarchical geometric models for visible surface algorithms. *Commun. ACM* 19 (10): 547–554.

29 N. Beckmann, H. Kriegel, R. Schneider. et al. The r*-tree: An efficient and robust access method for points and rectangles. Proc. SIGMOD Conf. on Management of Data, pages 322–331, 1990.

30 S. Gottschalk, M. Lin, and D. Manocha. Obb-tree: A hierarchical structure for rapid interference detection. In Proc. of ACM Siggraph'96, pages 171–180, 1996.

31 Bai L, Chang C, Wang Y. OBB Intersect Test Algorithm Based on Effective Constraint[C]. Proceedings of The fourth International Conference on Information Science and Cloud Computing (ISCC2015). Guangzhou: 2015.

32 Hubbard, P.M. (1995). Collision detection for interactive graphics applications. *IEEE Trans. Visual. Comput. Graphics* 1 (3): 218–230.

33 Klosowski, J., Held, M., Mitchell, J.S.B. et al. (1998). Efficient collision detection using bounding volume hierarchies of k-DOPs. *IEEE Trans. Visual. Comput. Graphics* 4 (1): 21–36.

34 E. Larsen, S. Gottschalk, M. Lin, et al. Fast proximity queries with swept sphere volumes. Technical Report TR99-018, Department of Computer Science, University of North Carolina, 1999.

35 S. Ehmann and M. Lin. Accurate and fast proximity queries between polyhedral using convex surface decomposition. In Proc. Eurographics, 2001.

36 Rosenberg L B. Virtual fixtures: Perceptual tools for telerobotic manipulation[C]//Virtual Reality Annual International Symposium, 1993., 1993 IEEE. IEEE, 1993: 76–82.

37 Bettini, A., Marayong, P., Lang, S. et al. (2004). Vision-assisted control for manipulation using virtual fixtures. *IEEE Trans. Robot.* 20: 953–966.

38 Park S, Howe RD, Torchiana DF. Virtual fixtures for robotic cardiac surgery. In: Proceedings of the 4th international conference on medical image computing and computer-assisted intervention. London (UK): Springer; 2001. p. 1419–1420.

39 Abbott, J.J., Marayong, P., and Okamura, A.M. (2007). Haptic virtual fixtures for robot-assisted manipulation. *Robot. Res.* 28: 49–64.

40 Passenberg, C., Peer, A., and Buss, M. (2010). A survey of environment-, operator-, and task-adapted controllers for teleoperation systems. *Mechatronics* 20: 787–801.

41 Oussama, K. (1986). Real-time obstacle avoidance for manipulators and mobile robots. *Int. J. Robot. Res.* 5 (1): 90–98.

42 Ge, S.S. and Cui, Y.J. (2002). Dynamic motion planning for mobile robots using potential field method. *Auton. Robots* 13: 207–222.

43 Poty, A., Melchior, P., Oustaloup, A.. Dynamic path planning for mobile robots using fractional potential field. In: Proceedings of the first international symposium on control, communications and signal; 2004. p. 557–561.

44 Alessandro Renzaglia, Agostino Martinelli. Potential field based approach for coordinate exploration with a multi-robot team. In: 8th IEEE international workshop on safety, security and rescue, robotics; 2010. p. 1–6.

45 Ni, T., Zhang, H.Y., Xu, P. et al. (2013). Vision-based virtual force guidance for tele-robotic system. *Comput. Electr. Eng.* 39 (7): 2135–2144.

46 F. Barbagli, K. Salisbury, and R. Devengenzo, Enabling Multi-finger, Multihand Virtualized Grasping, Proc. IEEE Int'l Conf. Robotics and Automation (ICRA 03), vol. 1, IEEE CS Press, 2003, pp. 806–815.

47 C. Basdogan, C.H. Ho, and M.A. Srinivasan, "A Ray-Based Haptic Rendering Technique for Displaying Shape And Texture of 3D Objects in Virtual Environments," Proc. ASME Dynamic Systems and Control Division, vol. 61, ASME, 1997, pp. 77–84.

48 Yi, L., Zhang, Y., Ye, X. et al. (2016). Haptic rendering method based on generalized penetration depth computation. *Signal Process.* 120: 714–720.

49 C.B. Zilles, J.K. Salisbury, A constraint-based god-object method for haptic display, in: Proceedings of the IEEE/RSJ International Conference on Intelligent Robots and Systems, Pittsburgh, 1995, pp. 146–151.

50 W. Mcneely, K. Puterbaugh, J. Troy, Six degree-of-freedom haptic rendering using voxel sampling, in: Proceedings of the ACM SIG-GRAPH, ACM Press, 1999, pp. 401–408.

51 M.A. Otaduy, M.C. Lin, Sensation preserving simplification for haptic rendering, in: Proceedings of the ACM Transaction on Graphics (Proceedings of ACM SIGGRAPH), 2003, pp. 543–553.

52 M. Ortega, S. Redon, S. Coquillart, A six degree-of-freedom godobject method for haptic display of rigid bodies, in: Proceedings of the IEEE Conference on Virtual Reality, 2006, pp. 191–198.

53 Arthur Gregory, Ajith Mascarenhas, Stephen Ehmann et al.Six Degree-of-Freedom Haptic Display of Polygonal Models. Proceedings of the conference on visualization '00, 10/2000

54 Kim, Y.J., Otaduy, M.A., Lin, M.C. et al. (2003). Six-degree-of -freedom haptic rendering using incremental and localized computations. *Presence* 12 (3): 277–295.

55 Moustakas, K. (2016). 6DoF haptic rendering using distance maps over implicit representations. *Multimedia Tools Appl.* 75: 4543–4557.

56 Miguel Angel Otaduy Tristan. 6-DOF haptic Rendering Using Contact Levels of Detail and Haptic Textures. Dissertation of the University of North Carolina at Chapel Hill, 2004

57 Jafari, A. and Ryu, J.-H. (2015). 6-DOF extension of memory-based passivation approach for stable haptic interaction. *Intell. Serv. Robot.* 8: 23–34.

58 Wagner, C.R., Lederman, S.J., Howe R.D.: A Tactile Shape Display Using RC Servomotors. In: 10th Symposium on Haptic Interfaces for Virtual Environment and Teleoperator Systems, Orlando, FL, March (2002)

59 Fukuda, T., Morita, H., Arai, F., et al. Micro Resonator Using Electromagnetic Actuator for Tactile Display. MHS, Nagoya, Japan (1997)

60 Taylor, P.M., Moser, A., and Creed, A. (1998). *A Sixty-four Element Tactile Display Using Shape Memory Alloy Wires*, 163–168. Amsterdam: Elsevier Science.

61 Haga, Y., Mizushima, M., Matsunaga, T. et al. (2005). Medical and welfare applications of shape memory alloy microcoil actuators. *Smart Mater. Struct.* 14 (5): S266–S272.

62 Pasquero, J., Hayward, V.: STReSS: A Practical Tactile Display System with One Millimeter Spatial Resolution and 700 Hz Refresh Rate. Eurohaptics, pp. 94–110, Dublin, Ireland (2003)

63 Kenaley, G.L., Cutkosky, M.R.: Electrorheological Fluid-Based Robotic Fingers With Tactile Sensing. Proc. In: IEEE International Conference on Robotics and Automation, Scottsdale, AR, pp. 132–136 (1989)

64 Tan, Hong Z; Pentland, Alex. Tactual displays for sensory substitution and wearable computers. International Conference on Computer Graphics and Interactive Techniques: ACM SIGGRAPH 2005 Courses: Los Angeles, California; 31 July–04 Aug. 2005, 07/2005.

65 Benali Khoudja, M., Hafez, M., Alexandre, J.M. et al. The VITAL Interface: A Vibrotactile Interface for Blind Persons. In: IEEE International Conference on Robotics and Automation (ICRA), New Orleans, LA, April 26–May 1 (2004)

66 Vujic, N., Hafez, M., Boy, P.: Thick Film Deposition of Piezoceramics Using Sol Gel Technology, Application to a Tactile Interface. In: 9th International Conference on New Actuators, ATUATOR 2004, Bremen, Germany, 14–16 June (2004)

67 Penn P, Petrie H, Colwell C, et al.. The haptic perception of texture in Virtual environments: an investigation with two devices. Proceedings of the First International Workshop on Haptic Human Computer Interaction .Berlin, 2001:25–30.

68 Childs, T.H.C. and Henson, B. (2007). Human tactile perception of screen-printed surfaces: self-report and contact mechanics experiments. *Proc. Inst. Mech. Eng. Part J J. Eng. Tribol.* 221: 427–441.

69 Derler, S., Gerhardt, L.C., Lenz, A. et al. (2009). Friction of human skin against smooth and rough glass as a function of the contact pressure. *Tribol. Int.* 42: 1565.

70 Goodwin, A.W., Macefield, V.G., and Bisley, J.W. (1997). Encoding of object curvature by tactile afferents from human fingers. *J. Neurophysiol.* 78 (78): 2881–2888.

71 Jemnalm, P., Birznieks, I., Goodwin, A.W. et al. (2003). Influence of object shape on responses of human tactile afferents under conditions characteristic of manipulation. *Eur. J. Neurosci.* 18 (1): 164–176.

72 Skedung, L., Arvidsson, M., Chung, J.Y. et al. (2013). Feeling small: exploring the tactile perception limits. *Sci. Rep.* 3.

73 Lisa Skedung. Tactile Perception: Role of Friction and Texture. Dissertation of Royal Institute of Technology. 2012

74 D.C. Ruspini, K. Kolarov, and O. Khatib, The Haptic Display of Complex Graphical Environments, Proc. ACM Siggraph, ACM Press, 1997, 345–352.

75 Hayward, V. and Armstrong, B. (2000). A new computational model of friction applied to haptic rendering. In: *Experimental Robotics VI* (ed. P. Corke and J. Trevelyan), LNCIS 250, 403–412. Springer-Verlag.

76 M. Minsky, Computational Haptics: The Sandpaper System for Synthesizing Texture for a Force Feedback Display, doctoral dissertation, Mass. Inst. of Technology, 1995.

77 Costa, M. and Cutkosky, M. (2000). Roughness Perception of Haptically Displayed Fractal Surfaces. *Proc. ASME Dynamic Systems and Control Division* 69 (2): 1073–1079.

78 J. Siira and D. Pai, Haptic Textures: A Stochastic Approach, Proc. IEEE Int'l Conf. Robotics and Automation (ICRA96), IEEE CS Press, 1996, pp. 557–562.

79 Derler, S. and Gerhardt, L.C. (2011). Tribology of skin: review and analysis of experimental results for the friction coefficient of human skin. *Tribol. Lett.* 1–27.

80 Masen, M.A. (2011). A systems based experimental approach to tactile friction. *J. Mech. Behav. Biomed. Mater.* 4: 1620–1626.

81 Wiertlewski, M., D. Leonardis, D. J. Meyer et al. A High-Fidelity Surface-Haptic Device for Texture Rendering on Bare Finger, Eurohaptics 2014: Springer, pp. 241–248, 2014.

82 Bergmann-Tiest, W.M. and Kappers, A.M.L. (2006). Analysis of haptic perception of materials by multidimensional scaling and physical measurements of roughness and compressibility. *Acta Psychol.* 121 (1): 1–20.

83 Smith, A.M., Chapman, C.E., Deslandes, M. et al. (2002). Role of friction and tangential force variation in the subjective scaling of tactile roughness. *Exp. Brain Res.* 144 (2): 211–223.

84 Bensmaia, S.J. and Hollins, M. (2003). The vibrations of texture. *Somatosens. Motor Res.* 20 (1): 33–43.

85 Wiertlewski, M., Lozada, J., and Hayward, V. (2011). The spatial spectrum of tangential skin displacement can encode tactual texture. *IEEE Trans. Robot.* 27 (3): 461–472.

86 Juan, W., Ju, Y., Li-yuan, L. et al. (2013). Design and implementation of measurement-based texture force rendering. *J. Syst. Simul.* 25 (11): 2630–2636.

87 Vasudevan H, Manivannan M. Recordable haptic textures Proceedings of IEEE International Workshop on Haptic Audio Visual Environments and their Applications. Ottawa, Canada: IEEE, 2006: 130–133.

88 Lang, J. and Andrews, S. (2011). Measurement-based modeling of contact forces and textures for haptic rendering. *IEEE Trans. Visual. Comput. Graphics* 17 (3): 380–391.

89 Okamura A, Dennerlein J, Howe R. Vibration feedback models for virtual environments [C]// Proceedings IEEE International Conference Robotics and Automation. Leuven, Belgium: IEEE, 1998: 674–679

90 Guruswamy, V.L., Lang, J., and Lee, W. (2011). IIR filter models of haptic vibration texture. *IEEE Trans. Instrum. Meas.* 60 (1): 93–103.

91 Craig G. McDonald and Katherine J. Kuchenbecker. Dynamic simulation of tool-mediated texture interaction. In Proc. IEEE World Haptics Conference, pp. 307–312. Daejeon, South Korea, 2013.

92 Shi, Y, Pai, D.K. Haptic display of visual images. Virtual Reality Annual International Symposium, 1997: 188–191

93 Tian, L., Song, A., and Chen, D. (2017). Image-based haptic display via a novel pen-shaped haptic device on touch screens. *Multimedia Tools Appl.* 76: 14969–14992.

94 Han, X.G., Ji-Ting, L.I., and Wang, D.X. (2011). Study on contour force rendering technology oriented to image perception. *J. Syst. Simul.* 23 (4): 713–718.

95 Li, J., Song, A., Wu, J. et al. (2010). Research on the method of haptic texture display based on SFS. *Chin. J. Sci. Instrum.* 31 (4): 812–817.

96 Rasool S, Sourin A. Image-driven haptic rendering, Transactions on computational science XXIII. Springer Berlin, Heidelberg, 2014. pp 58–77

97 Juan Wu. Research on Improving the Haptic Texture Display Using Stochastic Noise. 2009 IEEE International Conference on Robotics and Biomimetics (ROBIO). Pages:2125–2129

98 S. Xu, C. Li, L. Hu et al. An improved switching vector median filter for image-based haptic texture generation, 2012 5th International Congress on Image and Signal Processing, Chongqing, Sichuan, China, 2012, pp. 1195–1199.

99 Vidal-Verdu, F. and Hafez, M. (2007). Graphical tactile displays for visually-impaired people. *IEEE Trans. Neural Syst. Rehabil. Eng.* 15 (1): 119–130.

100 Bolanowski, S.J. Jr., Gescheider, G.A., Verrillo, R.T. et al. (1988). Four channels mediate the mechanical aspects of touch. *J. Acoust. Soc. Am.* 84: 1680.

101 Okamoto, S., Konyo, M., Saga, S. et al. (2009). Detectability and perceptual consequences of delayed feedback in a vibrotactile texture display. *IEEE Trans. Haptics* 2 (2): 73–84.

102 Smith, A.M., Gosselin, G., and Houde, B. (2002). Deployment of fingertip forces in tactile exploration. *Exp. Brain Res.* 147 (2): 209–218.

103 Miyaoka, T., Mano, T., and Ohka, M. (1999). Mechanisms of fine-surface-texture discrimination in human tactile sensation. *J. Acoust. Soc. Am.* 105: 2485–2492.

104 Hollins, M. and Risner, S.R. (2000). Evidence for the duplex theory of tactile texture perception. *Percept. Psychophys.* 62: 695–705.

105 Millet, G., Haliyo, S., Regnier, S., et al. The ultimate haptic device: First step. In: World Haptics Conference, IEEE (2009) 273–278

106 Verrillo, R.T. (1963). Effect of contactor area on the vibrotactile threshold. *J. Acoust. Soc. Am.* 35: 1962.

107 Wiertlewski, M. and Hayward, V. (2012). Mechanical behavior of the fingertip in the range of frequencies and displacements relevant to touch. *J. Biomech.* 45 (11): 1869–1874.

6

Virtual Simulation of Robot Control

6.1 Overview of Robot Simulation

Robot system simulation is to simulate the actual robot system by computer. It is mainly used to conduct a virtual test of the robot control algorithm. As a computer-aided tool, the simulation of robot system evolves closely with the development technology of robot.

Robot simulation focuses on three-dimensional geometric modeling of manipulator entities [1–4], analysis of robot kinematics and dynamics [5–9], trajectory and path planning [10–12], robot interaction with the working environment [13, 14] and humans [14, 15], and offline programming [16–19]. Furthermore, much research has focused on application simulation in special fields, such as robotic surgery [20, 21], robotic welding [22, 23], and massive multi-robot system [24].

Many robot simulation packages and platforms are also available, including [25–27] MRDS (Microsoft Robotics Developer Studio) (https://msdn.microsoft.com/en-us/library/bb483024.aspx) [28], which uses NVIDIA™ PhysX™ Technology to enable real-world physics simulation for robot models; Adams (http://www.mscsoftware .com/product/adams), which is very strong on dynamics simulation to support robot kinematics and dynamics simulation; Webotsk [29, 30], a commercial program for the simulation and prototyping of mobile robots, developed and supported by Cyberbotics Ltd., a leading company in simulation software founded in 1998 as a spinoff from the Swiss Federal Institute of Technology in Lausanne (EPFL) [31, 32]; Matlab, which has the widely used Robotics toolbox [33]; and Robot Operating System (ROS) (http://www.ros.org), which provides hardware-driven interfaces for commonly used robots and sensors. Famous process simulation software Delmia, as shown in

Dynamics and Control of Robotic Manipulators with Contact and Friction, First Edition.
Shiping Liu and Gang (Sheng) Chen.
© 2019 John Wiley & Sons Ltd. Published 2019 by John Wiley & Sons Ltd.

Figure 6.1 Industrial robot simulation module of DELMIA.

Figure 6.1, also has its own mainstream industrial robot simulation module. Tecnomatix® Robcad software enables the design, simulation, optimization, analysis, and offline programming of multidevice robotic and automated manufacturing processes in the context of product and production resources [34]. AutoMod is a famous discrete event simulation software, which can also be used to simulate robot systems (http://www.pmcorp.com/Portals/5/AutoMod%20Datasheet %20-%20Nov%202012.pdf). CimStation Robotics (CSR) is a comprehensive and easy-to-use robotic simulation tool available and works completely offline, eliminating the risk of damage to equipment and freeing robots and other equipment for round-the-clock production (www.acel.co.uk/cimstation-robotics).

Robot simulation systems are mainly based on desktop computers and workstations; however, there are also many based on the embedded platform, such as the one developed by Huazhong University of Science and Technology, based on the Tablet PC platform Huazhong NC robot offline programming system, shown in Figure 6.2.

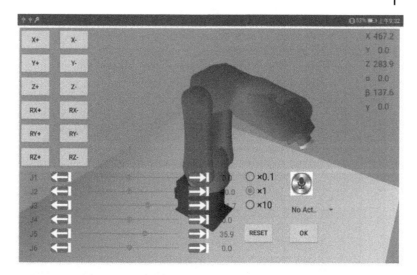

Figure 6.2 Off-line programming platform on Tablet PC.

General robot simulation mainly focuses on graphics and animation. Early robotics simulations were based primarily on the theory of multibody rigid body dynamics, focusing on efficient recursive algorithms [35]. If the goal is to make the simulation closer to the real world, we must pursue accurate physics simulation. Accurate contact modeling is now a hot topic for research [35–39].

6.2 3D Graphic Environment

Robot simulation generally needs three-dimensional graphic display to show geometric shapes and relative positions of the working environment and the robot itself. There are three main ways for this:

1) Direct use of graphic process function and mechanism analysis module of commercial 3D CAD software [17, 40, 41]. Therefore, the three-dimensional solid data structure, definition method of constraints between components, modeling methods, etc., are used directly. 3D CAD software always provides the application program interface (API) for developers working on robot simulation programs.

2) Using the graphics development kits to develop the graphics display system [42, 43]. Users need to define their own graphic data structure, parameter-driven relationship, and the connection and constraint relationship between the entities. The advantage is that the developed simulation system doesn't need to install other commercial 3D CAD systems in advance. Typical graphics development kits include Open Inventor, Java3D, Chai3D, Coin3D, Unity, PhysX, Open Scene Graph, and ARToolKit. Some can also provide 3D physics simulation functions [35], such as Chai3D, Bullet, Havok, MuJoco, ODE, PhysX [35], and Open Scene Graph.

3) Direct use of operating system or simulation software oriented only for robot applications. These systems are mentioned in Section 6.1. Researchers or developers use the graphic functions or modules in these systems directly. Generally, the bottom layer of these commercial robotic simulation systems is still based on OpenGL to achieve their graphic display and process functions. Some simulation systems also have haptic devices to enable human-computer interaction in a virtual environment [44].

In graphical simulation, collisions should be detected and avoiding strategies should be constructed. Sometimes, collision response should be simulated using multibody dynamics system theories. There is much research on this topic. Section 5.3.1 discussed this problem for virtual force calculation. These algorithms and strategies are also directly applied in graphical simulation. Some commercial robot simulators provide these functions directly.

6.3 Virtual Reality—Based Robot Control

6.3.1 Overview of Virtual Reality

Ivan Sutherland first proposed virtual reality (VR) in 1965, and over the next few years built a working system [45, 46]. In the 1990s, there were already many VR demos. Currently, many VR-based commercial entertainment, vehicle manipulation simulators, and aircraft manipulation simulators are sold and work well [47–49]. Generally, a VR system needs to achieve three features: interactivity, illusion of immersion, and imagination (3I features). Modeling, presentation, and interaction are three key technologies in building a virtual system [50].

VR modeling refers to modeling of 3D objects, their interaction with virtual and real world, and the effect of the interaction on the objects, etc., in order to simulate them [51]. Typical methods can be divided into these categories: scene appearance modeling method, behavior modeling method, modeling of virtual-real combination, and physical based modeling method [52]. VR presentation technology is to provide the user a realistic visual insight, sense of hearing, feeling, etc. The presentation technologies of VR are divided into these categories: auditory presentation technology, tactility presentation technology, and visual presentation technology [51]. VR human-machine interaction is essential for users to obtain a vivid perception by operating various virtual objects in a virtual environment, which mainly involves information exchange modes and equipment of interaction and mutual influence between people and the virtual environment [50]. Generally, these interaction modes or equipments are used:

- Scene display mode and equipment including head-mounted display (HMD), projection display like CAVE, hand-held display, etc.
- Force/tactility interaction modes and equipment
- Tracking location mode and equipment
- Walking interaction modes and equipment

Several problems still inhibit development of VR technology and applications, such as presentation of more physics features and new physics models, soft and tough feeling, and other sensory channels [53].

VR has been used in many fields. NASA developed the Station Spacewalk Game to allow the user to experience the thrill of conducting NASA repair work on the International Space Station. This video game features simulations of actual Extravehicular Activities (EVAs) conducted by NASA astronauts on missions to provide power to the space station [49].

6.3.2 Overview of Teleoperation

Teleoperation is a robotics application wherein a master and a slave system interact with each other and the environment [54–56]. It is mainly used to control the remote robot or robot working in an environment where human presence is prohibited or inconvenient, such as a space station manipulator, lunar rover, underwater robot, surgical robot, firefighting robot, robots working inside nuclear power stations, etc. Generally, in teleoperation there is feedback information sensed by the teleoperator (robot), such as image or video information, sound information, or measurement data of physics variants (e.g., force, velocity, acceleration, position, temperature, etc.).

In general, a classical teleoperation system consists of five interconnected elements: a human operator, man-machine interface, remotely located teleoperator, surrounding environment, and the communication channel [57–59].

Generally, there are three control modes of teleoperations: direct control, supervisory mode, and autonomous mode. In direct mode, the operator controls the remote robot directly with a hand-controller or by mouse and keyboard, while watching video from cameras mounted on a remote robot (i.e., slave robot) [56, 60]. If there is a significant time delay in the communication channel, the strategy of move-and-wait [61] is used. In the supervisory mode [62, 63], the remote robot has certain autonomous intelligence and operators intervene with the remote robot only in special cases, such as when errors occur in the actions of the remote robot, or if it can't complete a complex task. In autonomous mode, the remote robot has enough intelligence to complete its task, so there is no need for the operator to intervene. Generally, it is difficult to achieve a robot with very high autonomous intelligence. Direct mode always takes too much time and has very low efficiency. So, supervisory mode is widely used in many teleoperation applications.

One of the most important features of teleoperation is telepresence. If users receive sufficient information about the slave robot and the remote environment they are interacting with, they feel as they are actually present at the remote site. This condition is commonly referred to as telepresence [64, 65]. Currently, telepresence is mainly achieved by VR means.

There are mainly two types of telepresence: visual telepresence, which depends on image (or video) information or 3D modeling of the remote environment; and haptic telepresence, which the operator senses through a haptic device. Although the slave-mounted camera can provide certain information of the working environment, the virtual environment is still important for the operator to plan in advance, to help visualize the entire space due to objects blocking the camera's view. Transparency describes the discrepancy between remote and local presence. Ideal transparency means that the user is not able to distinguish remote presence from local presence [66].

Typically, there are time delays in a teleoperation system. To improve operation efficiency, it needs to analyze properties of time delay, simulate time delay, and use prediction algorithms [67] to predict the current movement of the remote robot. At the operator's end, VR is generally used to improve the telepresence.

6.3.3 Virtual Reality—Based Teleoperation

VR is an important means to solve the large time-delay problem in teleoperation. At the operator's end, the virtual environment built from known knowledge of the remote environment can provide predictive display and make virtual validation of future operations of the remote (slave) robot. Sometimes operators can also program in this virtual environment to create command packages for the remote robot to execute continuously for a certain amount of time to improve efficiency. Unfortunately, an exact match between the virtual world and the real world can never be guaranteed [68].

Figure 6.3 shows the 3D environment of a space robot. This virtual environment is built by the graphics toolkit Open Inventor [69], developed by Thermo Fisher Scientific Company. ProE is also used to

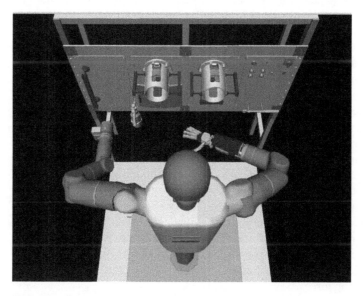

Figure 6.3 The 3D environment of a space robot.

build complex 3D models of some objects, which are input into Open Inventor. In this virtual environment, any structural features inside each object are deleted, and only the shell (appearance) is retained. The connections and kinematic relationships between objects or links of the robot are defined by node trees in Open Inventor.

Some typical VR devices, such as the HMD device, haptic device, and data glove are used in the VR-based teleportation system [70]. HMD devices or 3D glasses are used to present the virtual environment or the augmented site scene to give the visual telepresence. The haptic device (e.g. a hand- controller (HC) with force feedback) is used to present the haptic telepresence. The data glove is mainly used to control a dexterous robotic hand. Some types of data gloves also have a force feedback function.

In this virtual environment, virtual fixtures can also be applied. The virtual fixture was conceived by Rosenberg [71]. Generally, the operator manipulated the hand-controller, which constantly has the force feedback function to move the virtual robot or remote robot. The virtual fixture can offer assistance such as avoiding collision, reducing the pressure or the tremor of the operator, achieving the accurate movement of robot, and approaching the end of manipulator rapidly to specified places, etc., by guiding the operator along a task-specific pathway or limiting the operator to stay within a "safe" region, which could be on a surface or in a volume, or forbidding them from entering into certain areas or volumes. Typical constraints include the constraints of point, line, curve, plane, curved surfaces, volume area, value set of coordinates, and so on [72]. Sometimes, these constraints can also be dynamic, which means that these virtual fixtures can change with time or the movement of the end of the robot with respect to the working environment.

Virtual fixtures confine the movement of the end effecter of the virtual or remote robot, which is manipulated by the operator. If the operator uses a haptic device to control the movement of the robot, the force that the operator feels on the haptic device will lead toward or away from the ideal trajectory or specified region. The force will increase with the increasing deviation degree that the end of the robot deviates from the ideal the trajectory or increase rapidly and greatly when the robot end approaches the forbidden region. Generally, the artificial potential field method is used to calculate these guiding or resisting forces.

6.4   Augmented Reality—Based Teleoperation

6.4.1   Overview of Augmented Reality

Augmented reality (AR) is a technology to superimpose different types of information, such as text, signs, or graphical scenes (virtual objects), etc., which are created by computers, over a real environment or a real-time video scene collected from the real environment. Generally, virtual objects should blend into the real-world scene seamlessly; furthermore, this blending should change with the changes in the real world scene in context. In VR, the real environment is modeled and simulated. Generally, it is difficult to model the real world accurately and completely. It always needs great effort and much time on building the virtual environment. Sometimes it is even impossible. In AR, there is no need to build the complete virtual environment, and the virtual information enriches the surrounding environment. In Figure 6.4, the real object can also be processed to be transparent to show the internal colored virtual objects indicating the position for the real components to be assembled [73].

Generally, AR uses these display technologies, such as HMD, handheld displays (using PDA, smart phone, Tablet PC, etc.), and spatial displays (spatial augmented reality (SAR) making use of video-projectors, optical elements, holograms, radio frequency tags, and other tracking technologies to display graphical information directly onto physical objects without requiring the user to wear or carry the display [74, 75].).

(a) The real working environment   (b) Augmented reality scene
and the 3 DOF robot

Figure 6.4 One example of augmented reality application.

AR can potentially apply to all other senses besides sight, augmenting smell, touch, and hearing as well. AR can also be used to augment or substitute for users' missing senses by sensory substitution, such as augmenting the sight of blind users or users with poor vision by the use of audio cues, or augmenting hearing for deaf users by the use of visual cues [75–77].

In AR systems, generally there are three coordinate frames: the camera coordinate frame, the real-world coordinate frame, and the virtual object coordinate frame. To blend the virtual objects with the real ones, the transformation relationship among these three coordinate frames must be calculated to achieve the registration (alignment).

6.4.2 Augmented Reality–Based Teleoperation

In a teleoperation system, HC is sometimes used to control the slave robot in the remote environment. Sometimes, there are misalignments of the coordinates of the end effector axes and the HC axes, which hinders operator performance. Using AR, the coordinate system of the end effector can be graphically overlaid in three dimensions on the video views with uniquely colored axes [78]. The same color scheme is used to label the corresponding axes on the HCs. Operators use these color cues to map each axis on the HCs to the corresponding colored axis of the augmented coordinates at the end effector to obtain end effector movement in the desired direction [78]. This benefits the operator greatly due to the intuitive mode and reduction of mental workload.

In some teleoperation systems with significant time delay, these virtual objects can also be a predictive display to show the expected current position of the robot because the video scene shown at the operator side is the state of the remote environment at a previous time, not the current state. During operation, the augmented coordinates precedes the robot's movements due to time delay; these can also be considered as the predictive display for the end effector.

Virtual fixture can also be used in AR teleoperation systems. In literature [79] a virtual barrier that the robot is not supposed (and able) to cross to avoid collisions is visualized in an AR-view of the surroundings. The guiding path or forbidden region can also be reconstructed from real-time onsite images [80–82].

6.5 Task Planning Methods in Virtual Environment

6.5.1 Overview

The task planning method has a direct impact on the complexity of the planning process and the accuracy of the planning results. Therefore, the task planning method must still be explored, and a unified method has not yet been formed.

Rybski et al. [83] proposed a method of planning for mobile robots to learn how to accomplish various tasks autonomously by observing human behavior and listening to people's speeches. This method only combines the speech recognition technology, and the planning ability is very limited; the robot can carry out the task that is completely based on the navigation path search, but this method fails to do complex task planning. Mikita et al. [84] proposed a method of task planning for service robots in a partially unknown real mission environment, which allows the robot to obtain unknown information about the environment through human–computer interaction, and describes the behavior and tasks of the robot using the Planning Domain Definition Language (PDDL). The method cannot realize the real-time planning of the task; the user can only provide the robot with unknown environmental information and cannot interfere with the implementation of the task process. Kaelbling and Lozano-Pérez [85] classify tasks and implement real-time online planning, but this approach relies heavily on the accuracy of hierarchical partitioning and must follow the top-down selection model, with great limitations. Karlsson et al. [86] integrated mission planning and path planning in a humanoid robot system, taking geometrical backtracking and validating on Justin robots. Tzafestas and Plouzennec [87] implemented an expert prototype system for planning robotic automatic assembly processes, but only applied in the field of assembly, with a very limited range of applications. Based on the Petri net, Costelha and Lima [88] have established a framework that covers the task modeling, analysis, and execution of a robot, but it can only be modeled and simulated for specific tasks and is not universal. Birkenkampf, Leidner et al. argue that it is challenging to achieve completely autonomous task planning in an unstructured and changing environment, but a possible coping strategy is autonomy with supervision, in which the operator remotely controls the robot at a highly abstract level

[89, 90]. Although this approach can balance the workload between man and machine, the planning efficiency will drop rapidly as the task scene becomes more complex and the number of operational objects increases dramatically.

6.5.2 Interactive Graphic Mode

While remotely operating a robot, the focus of the operator should be on the task to be solved, rather than on the robot to be operated. This can be supported by a human–robot interaction (HRI) replicating the real-world experience, as if someone would solve the given task on its own. In order to achieve this, the robot has to have the same information about the environment and the contained objects as the operator would have in its place. This issue can be addressed by facilitating an object-centered knowledge base that provides symbolic and geometric object information to the robot. With this information, it is possible to generate actual feasible action possibilities from the current world state. To avoid distractions from too much information while still guiding the operator's decisions, an intuitive user interface (UI) design along with a straightforward command concept is necessary. Several intuitive graphic methods, such as composing these predefined motion packages, controlling virtual individual joint directly, etc. to plan the behaviors of the humanoid robot toy is provided for the NAO humanoid robot (https://www.ald .softbankrobotics.com/en/robots/nao). Figure 6.5 illustrates another

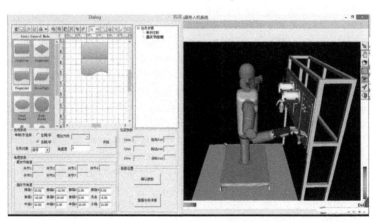

Figure 6.5 Task planning in graphic interaction mode.

intuitive graphic interaction mode. It is used for the task planning in the teleoperation of a space robot.

Prior knowledge is provided for all known objects by the robot. The objects are hierarchically arranged in the object-oriented paradigm and categorized by functionality. Objects of the same class share the same process models for handling and can therefore be manipulated in the same way while considering their specific properties such as size and shape. The world representation holds the current state of the environment of the robot. Objects are instantiated here with specific symbolic and geometric properties.

The task reasoning is performed in the context of the objects related to the task. The object functionality is therefore stored in so-called action templates that define distinct manipulation instructions. Those action templates consist of two segments, which are evaluated in a two-step hybrid reasoning approach. First, the symbolic headers defined in the PDDL [91] are parsed in order to construct the symbolic domain and to solve a given task symbolically. The resulting symbolic transition is evaluated in the second step. Therefore, the geometric body of the action templates is evaluated to ground the symbolic actions into robot-specific actions using modular geometric simulations such as navigation, motion planning, or dynamics simulations. Geometric backtracking in case of unsuccessful simulation is inherent in this step [92]. This approach has been successfully evaluated to solve everyday manipulation tasks [12], mobile manipulation tasks [93], and force-sensitive whole-body manipulation [94].

The method proposed above generates large action sets for complex world states. However, presenting all of these actions to the operator is unfeasible, as it results in high cognitive load. Therefore, an UI concept based on a point-and-click paradigm and object-centered information reduction has been developed.

In a nutshell, the point-and-click paradigm defines a model whereby pointing to a specific location indicates user interest and clicking executes some kind of action. This approach is commonly used in Mac- or Windows-based operating systems and adventure video games. The latter often facilitates control over a virtual agent whose movements are directed but not completely specified by the user, by clicking on objects in the virtual world. By this, high-level targets are specified that are reasonable for the clicked object in the current context. The corresponding actions to achieve these targets are autonomously scheduled and executed by the agent.

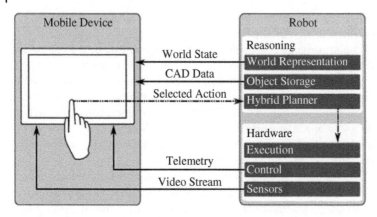

Figure 6.6 The interaction between the mobile device and the robot.

A prototype of the proposed HRI is implemented for tablet computers. The complexity of the system is distributed between the mobile device and the robot, as shown in Figure 6.6. Modules for reasoning, as previously described, and hardware access are provided by the robot. The mobile device implements the visualization and the handling of operator input, as previously described.

References

1 Sheng, X. and Hirsch, B.E. (1991). Geometric modeling for robot simulation. *Robotersysteme* 7 (2): 85–90.

2 Seo, J.W., Haas, C., and Saidi, K. (2007). Graphical modeling and simulation for design and control of a tele-operated clinker clearing robot. *Autom. Constr.* 16: 96–106.

3 Bouzgou, K. and Ahmed-Foitih, Z. (2014). Geometric modeling and singularity of 6 DOF Fanuc 200IC robot. Fourth edition of the International Conference on the Innovative Computing Technology (INTECH 2014), Luton, 208–214.

4 Morgansen, K.A., Triplett, B.I., and Klein, D.J. (2007). Geometric methods for modeling and control of free-swimming fin-actuated underwater vehicles. *IEEE Trans. Robot.* 23 (6): 1184–1199.

5 Fueanggan, S. and Chokchaitam, S. (2009). Dynamics and Kinematics Simulation for Robots. International Association of Computer Science and Information Technology - Spring Conference, Singapore, 136–140.

6 Bouzgou, K. Amar, R.H.E. and Ahmed-Foitih, Z. (2015). Virtual reality simulation and singularity analysis of 3-RRR translational parallel robot, Fifth International Conference on the Innovative Computing Technology (INTECH 2015), Galcia, 61–66.

7 Sasiadek, J.Z. (1985). Simulation of kinematics and dynamics of robots using a symbolic manipulation system. SIAM Conference on Geometric Modeling and Robotics, Albany, NY (USA), 15–19 July.

8 Cretescu, N.R. Kinematic and dynamic simulation of a 3DOF parallel robot. *Bull. Transilvania Univ. Brasov.*. Engineering Sciences. Series I 8 (2): 73–78.

9 Aspragathos, N.A. and Dimitros, J.K. (1998). A comparative study of three methods for robot kinematics. *IEEE Trans. Syst.* 28.

10 Sud, A., Andersen, E., Curtis, S. et al. (2008). Real-time path planning in dynamic virtual environments using multiagent navigation graphs. *IEEE Trans. Visual Comput. Graphics* 14 (3): 526–538.

11 Mousavi, P.N., Nataraj, C., Bagheri, A. et al. (2008). Mathematical simulation of combined trajectory paths of a seven link biped robot. *Appl. Math. Modell.* 32 (7): 1445–1462.

12 Gasparetto, A. and Zanotto, V. (2008). A technique for time-jerk optimal planning of robot trajectories. *Rob. Comput. Integr. Manuf.* 24: 415–426.

13 Ikuta, K., Ishii, H., and Nokata, M. (2003). Safety evaluation method of design and control for human-care robots. *Int. J. Rob. Res.* 22 (5): 281–297.

14 Dombrowski, U., Stefanak, T., and Perret, J. (2017). Interactive simulation of human-robot collaboration using a force feedback device. 27th International Conference on Flexible Automation and Intelligent Manufacturing, FAIM 2017, 27-30 June 2017, Modena, Italy. *Procedia Manuf.* 11: 124–131.

15 Khatib, O., Brock, O., Chang, K.C. et al. (2003). Robots for the Human and Interactive Simulations. *Proceedings of the 11th World Congress in Mechanism and Machine Science*, August 18–21, Tianjin, China.

16 Schroer, B.J. and Teoh, W. (1986). A graphical simulation tool with off-line robot programming. *Simulation* 47 (2): 63–67.

17 Mitsi, S., Bouzakis, K.-D., Mansour, G. et al. (2005). Off-line programming of an industrial robot for manufacturing. *Int. J. Adv. Manuf. Technol.* 26 (3): 262–267.

18 Pan, Z., Polden, J., Larkin, N. et al. (2012). Recent progress on programming methods for industrial robots. *Rob. Comput. Integr. Manuf.* 28 (2): 87–94.

19 Hollingum, J. (1994). Simulation, calibration and off-line programming. *Ind. Rob.* 21 (5): 20–21.

20 Kunkler, K. (2006). The role of medical simulation: an overview. *Int. J. Med. Rob. Comput. Assisted Surg.* 2: 203–210.

21 Khan, Z.A., Kamal, N., Hameed, A. et al. (2017). Smart SIM – a virtual reality simulator for laparoscopy training using a generic physics engine. *Int. J. Med. Rob. Comput. Assisted Surg.* 13: e1771. doi: 10.1002/rcs.1771.

22 Fang, H.C., Ong, S.K., and Nee, A.Y.C. (2013). *Int. J. Adv. Manuf. Technol.* 67: 2033. doi: 10.1007/s00170-012-4629-7.

23 Chen, B. and Feng, J. (2014). Modeling of underwater wet welding process based on visual and arc sensor. *Ind. Rob.* 41 (3): 311–317. doi: 10.1108/IR-03-2014-0315.

24 Vaughan, R. (2008). Massively multi-robot simulation in stage. *Swarm Intell.* 2 (2–4): 189–208. doi: 10.1007/s11721-008-0014-4.

25 Koseeyaporn, P., Cook, G.E., and Strauss, A.M. (2002). Extendible Simulation Package for Robotics Research and Instruction. IASTED International Conference Modeling and Simulation (MS 2002), Marina Del Rey, California, USA, May 13–15.

26 Zha, X.F. and Du, H. (2001). Generation and simulation of robot trajectories in a virtual CAD-based off-line programming environment. *Int. J. Adv. Manuf. Technol.* 17 (8): 610–624.

27 Hale, J.G. Hohl B., Moraud, E.M. (2008) Robot simulation, collisions and contacts. https://www.researchgate.net/publication/228971493_Robot_simulation_collisions_and_contacts (accessed 14 May 2018).

28 Gates, B. (2006). A robot in every home D. *Sci. Am.* 28 (6): 13–21.

29 Cyberbotics Ltd. http://www.cyberbotics.com (accessed 14 May 2018)

30 Michel, O. (2004). Cyberbotics Ltd – Webotsk: Professional mobile robot simulation. *Int. J. Adv. Rob. Syst.* 1: 39–42.

31 Holh, L., Téllez, R.A., Michel, O. et al. (2006). Aibo and Webots: simulation, wireless remote control and controller transfer. *Rob. Auton. Syst.* 54 (6): 472–485.

32 Michel, O. (1998). Webots: Symbiosis between virtual and real mobile robots. In VW '98: Proceedings of the First International Conference on Virtual Worlds, (pp. 254–263). New York: Springer-Verlag.

33 Corke, P.I. A Computer tool for Simulation and Analysis the Robotics Toolbox for Matlab CSIRO Division of Manufacturing Technology http://www.petercorke.com/robot/ARA95.pdf

34 Robcad Robotics and automation workcell simulation, validation and off-line programming. https://www.plm.automation .siemens.com/en/products/tecnomatix/manufacturing-simulation/ robotics/robcad.shtml#lightview%26url=/en_us/Images/7777_ tcm1023-4965.pdf%26title=Robcad%26description=Robcad Fact Sheet%26docType=pdf (accessed 14 May 2018).

35 Erez, T., Tassa, Y., and Todorov, E. (2015). *Simulation Tools for Model-Based Robotics: Comparison of Bullet, Havok, MuJoCo, ODE and PhysX.* IEEE International Conference on Robotics and Automation (ICRA).

36 Baraff, D. (1994). Fast contact force computation for nonpenetrating rigid bodies," In Proceedings of the 21st annual conference on Computer graphics and interactive techniques. *ACM* 23–34.

37 Mirtich, B. and Canny, J. (1995). Impulse-based simulation of rigid bodies. In Proceedings of the 1995 symposium on Interactive 3D graphics. *ACM* 181–ff.

38 Stewart, D.E. and Trinkle, J.C. (1996). An implicit time-stepping scheme for rigid body dynamics with inelastic collisions and coulomb friction. *Int. J. Numer. Methods Eng.* 39 (15): 2673–2691.

39 Anitescu, M. and Potra, F.A. (1997). Formulating dynamic multi-rigid-body contact problems with friction as solvable linear complementarity problems. *Nonlinear Dyn.* 14 (3): 231–247.

40 Kamel, B., Amar, R.H.E., Zoubir, A.-F (2015). Virtual reality simulation and singularity analysis of 3-RRR translational parallel robot Proceedings of the Fifth International Conference on Innovative Computing Technology. INTECH: 61–66

41 Omura, Y. (2006). *Robotic software simulator*, vol. 61, 56–58. Design News.

42 Ruolong, Q.I., Weijia, Z., and Jinguo, L. (2013). XIAO Lei1An effective method for implementing virtual control and 3D simulation of robot motion in VC platform. *Robot* 35 (5): 594–599.

43 Park, J., Park, C.-H., Park, D.-i. et al. (2017). *The Library for Grasp Synthesis & Robot Simulation.* 14th International Conference on Ubiquitous Robots and Ambient Intelligence (URAI), 418–423. Korea: Maison Glad Jeju, Jeju.

44 Escobar-Castillejos, D., Noguez, J., Neri, L. et al. (2016). A review of simulators with haptic devices for medical training. *J. Med. Syst.* 40: 104. doi: 10.1007/s10916-016-0459-8.

45 Brooks, F.P. (1999). What's real about virtual reality? *IEEE Comput. Graphics Appl.* 19 (6): 16.

46 Sutherland, I. (1965). The ultimate display. In: *Multimedia: From Wagner to Virtual Reality* (ed. R. Packer and K. Jordan). New York: Norton.

47 Earnshaw, R.A. (ed.) (2014). *Virtual Reality Systems*. Cambridge, MA: Academic Press.

48 Rubio-Tamayo, J.L., Gertrudix Barrio, M., and García García, F. (2017). Immersive environments and virtual reality: systematic review and advances in communication, interaction and simulation. *Multimodal Technol. Interact.* 1: 21.

49 NASA. Station Spacewalk Game. https://www.nasa.gov/multimedia/ 3d resources/station spacewalk game.html. (accessed 14 May 2018)

50 Qin Ping, Z. (2009). A survey on virtual reality. *Sci. China*(Series F:Information Sciences) 52 (3): 348–400.

51 Singh, N. and Singh, S. (2017) Virtual reality: A brief survey. International Conference on Information Communication and Embedded Systems (ICICES), 2017:

52 Zhou, N.-N. and Deng, Y.-L. (2009). Virtual reality: a state-of-the-art survey. *Int. J. Autom. Comput.* 6 (4): 319–325.

53 Zhao, Q.P. (2011). 10 scientic problems in virtual reality. *Commun. ACM* 54: 116–118.

54 Ambrose, R.O., Aldridge, H., Askew, R.S. et al. (2000). Robonaut: NASA's space humanoid. *IEEE Intell. Syst. Appl.* 4: 57–63.

55 Dede, M. and Tosunoglu, S. (2006). Fault-tolerant teleoperation systems design. *Ind. Rob.* 33 (5): 365–372.

56 Fong, T. and Thorpe, C. (2001). Vehicle teleoperation interfaces. *Auton. Rob.* 11: 9. doi: 10.1023/A:1011295826834.

57 Chan, L., Naghdy, F., and Stirling, D. (2014). Application of adaptive controllers in Teleoperation systems: a survey. *IEEE Trans. Hum. Mach. Syst.* 44 (3): 337–352.

58 Hokayem, P.F. and Spong, M.W. (2006). Bilateral teleoperation: an historical survey. *Automatica* 42 (12): 2035–2057.

59 Anderson, R.J. and Spong, M.W. (1989). Bilateral control of teleoperators with time delay. *IEEE Trans. Autom. Control* 34 (5): 494–501.

60 Sheridan, T.B. and Ferrell, W.R. (1963). Remote manipulative control with transmission delay. *IEEE Trans. Hum. Factors Electron.* HFE-4 (1): 25–29.

61 Ferrell, W.R. (1965). Remote manipulative control with transmission delay. *IEEE Trans. Hum. Factors Electron.* HFE-6 (1): 24–32.

62 Ferrell, W.R. and Sheridan, T.B. (1967). Supervisory control of remote manipulation. *IEEE Spectr.* 4 (10): 81–88.

63 Sheridan, T.B. (1992). *Telerobotics, Automation, and Human Supervisory Control.* Cambridge: MIT Press.

64 Pacchierotti, C., Tirmizi, A., and Prattichizzo, D. (2014). Improving transparency in Teleoperation by means of cutaneous tactile force feedback. *ACM Trans. Appl. Percept.* 11 (1): Article 4): doi: 10.1145/2604969.

65 Draper, J.V., Kaber, D.B., and Usher, J.M. (1998). Telepresence. *Hum. Factors .J. Hum. Factors .Ergon. Soc.* 40 (3): 354–375.

66 Artigas, J., Ryu, J.-H., and Preusche, C. (2010). Time domain passivity control for position-position Teleoperation architectures. *Presence* 19 (5): 482–497.

67 Bemporad, A. (1998) Predictive control of teleoperated constrained systems with unbounded communication delays. In *Proc. IEEE International Conf. Decision and Control, Tampa*: 2133–2138.

68 Zainan, J., Hong, L., Jie, W. et al. (2009). Virtual reality-based teleoperation with robustness against modeling errors. *Chin. J. Aeronaut.* 22: 325–333.

69 Open Inventor™ | Open Inventor 3D SDK https://www.openinventor.com

70 Peer, A., Pongrac, H., and Buss, M. (2010). Influence of varied human movement control on task performance and feeling of Telepresence. *Presence* 19 (5): 463–481.

71 Rosenberg, L.B. (1993). Virtual fixtures: perceptual tools for telerobotic manipulation. *Proc. IEEE Virt. Real. Annual Int. Symp.* 76–82.

72 Bowyer, S.A., Davies, B.L., and y Baena, F.R. (2014). Active constraints/virtual fixtures: a survey. *IEEE Trans. Rob.* 30 (1): 138–157.

73 Chi, X. (2011). *Research on the 3D Objeet Registration Method in Augmented Reality and its Application.* Dissertation of Huazhong University of Science and Technology.

74 Bimber, O., Raskar R., and Inami, M. (2007). Spatial Augmented Reality. SIGGRAPH 2007 Course 17 Notes.

75 Furht, B. (2011). *Handbook of Augmented.* New York, Dordrecht, Heidelberg, and London: Springer Reality.

76 Azuma, R., Baillot, Y., Behringer, R. et al. (2001). Computer graphics and applications. *IEEE* 21 (6): 34–47.

77 Xiong, Y. (2005). Key Technology Research of Teleoperation Based on Augmented Reality Dissertation of Huazhong University of Science and Technology,

78 Chintamani, K., Cao, A., Darin Ellis, R. et al. (2010). Improved telemanipulator navigation during display-control misalignments using augmented reality cues. *IEEE Trans. Syst. Man Cybern. Syst. Hum.* 40 (1): 29–39.

79 Leutert, F. and Schilling, K. (2012). Support of Power Plant Telemaintenance with Robots by Augmented Reality Methods. 2nd International Conference on Applied Robotics for the Power Industry (CARPI) ETH Zurich, Switzerland. September 11–13, p 45–49.

80 Yamamoto1, T., Abolhassani, N., Jung, S. et al. (2012). Augmented reality and haptic interfaces for robot-assisted surgery. *Int. J. Med. Rob. Comput. Assisted Surg.* 8: 45–56.

81 Fusiello, A. and Murino, V. (2004). Augmented scene modeling and visualization by optical and acoustic sensor integration. *IEEE Trans. Visual Comput. Graphics* 10 (6): 625–636.

82 Portilla, H. and Basanez, L. (2007). Augmented reality tools for enhanced robotics teleoperation systems. 3DTV Conference, Kos Island): 1–4.

83 Rybski, P.E., Yoon, K., Stolarz, J., et al. (2007). Interactive robot task training through dialog and demonstration[C]//Human-Robot Interaction (HRI), 2nd ACM/IEEE International Conference on. IEEE, 49–56.

84 Mikita, H., Azuma, H., Kakiuchi, Y. et al. (2012). Interactive symbol generation of task planning for daily assistive robot[C]//Humanoid Robots (Humanoids), 12th IEEE-RAS International Conference on. IEEE, 698–703.

85 Kaelbling, L.P. and Lozano-Pérez, T. (2011). Hierarchical task and motion planning in the now[C]. Robotics and Automation (ICRA). *2011 IEEE International Conference on. IEEE*: 1470–1477.

86 Karlsson, L., Bidot, J., Lagriffoul, F. et al. (2012). Combining task and path planning for a humanoid two-arm robotic system[C]. *Proceedings of TAMPRA: Combining Task and Motion Planning for Real-World Applications (ICAPS workshop)*: 13–20.

87 Tzafestas, S.G. and Plouzennec, A.M. (2014). A Prolog-Based Expert System Prototype for Robot Task Planning[J]. Advanced Information Processing in Automatic Control (AIPAC'89): 225.

88 Costelha, H. and Lima, P. (2012). Robot task plan representation by petri nets: modelling, identification, analysis and execution[J]. *Auton. Rob.* 33 (4): 337–360.

89 Birkenkampf, P., Leidner, D., and Borst, C. (2014). A knowledge-driven shared autonomy human-robot interface for tablet computers[C]//Humanoid Robots (Humanoids), 14th IEEE-RAS International Conference on. IEEE: 152–159.

90 Leidner, D., Birkenkampf, P., Lii, N.Y, et al. (2014). Enhancing supervised autonomy for extraterrestrial applications by sharing knowledge between humans and robots[C]//Workshop on How to Make Best Use of a Human Supervisor for Semi-Autonomous Humanoid Operation, IEEE-RAS International Conference on Humanoid Robots (Humanoids).

91 Ghallab, M., Howe, A., Christianson, D. et al. (1998). PDDL-the planning domain definition language,. AIPS98 planning committee,. 78 (4): 1–27.

92 Leidner, D., Borst, C., and Hirzinger, G. (2012). Things are made for what they are: solving manipulation tasks by using functional object classes. *Proc. IEEE/RAS Int. Conf. Hum. Rob.* 429–435.

93 Leidner, D. and Borst, C. (2013 Hybrid reasoning for mobile manipulation based on object knowledge. In *Workshop on AI-based Robotics at IEEE/RSJ International Conference on Intelligent Robots and Systems (IROS).*

94 Leidner, D., Dietrich, A., Schmidt, F. et al. (2014). Object-centereyd hybrid reasoning for whole-body mobile manipulation. In *Proc. of the IEEE International Conference on Robotics and Automation (ICRA),* 1828–1835.

Index

Dynamics and Control of Robotic Manipulators with Contact and Friction, First Edition.
Shiping Liu and Gang (Sheng) Chen.
© 2019 John Wiley & Sons Ltd. Published 2019 by John Wiley & Sons Ltd.